DISCARDED

Problems of Coordination
in Economic Activity

RECENT ECONOMIC THOUGHT SERIES

Editor:

Warren G. Samuels
Michigan State University
East Lansing, Michigan, U.S.A.

Other books in the series:

Feiwel, G.:
 SAMUELSON AND NEOCLASSICAL ECONOMICS
Wade, L.:
 POLITICAL ECONOMY: MODERN VIEWS
Jarsulic, M.:
 MONEY AND MACRO POLICY
Samuelson, L.:
 MICROECONOMIC THEORY
Mirowski, P.:
 THE RECONSTRUCTION OF ECONOMIC THEORY
Lowry, Todd:
 PRE-CLASSICAL POLITICAL ECONOMY
Officer, L.:
 INTERNATIONAL ECONOMICS
Asimakopulos, A.:
 THEORIES OF INCOME DISTRIBUTION
Earl, P.:
 PSYCHOLOGICAL ECONOMICS; DEVELOPMENT, TENSIONS, PROSPECTS
Thweatt, W.:
 CLASSICAL POLITICAL ECONOMY
Peterson, W.:
 MARKET POWER AND THE ECONOMY
DeGregori, T.:
 DEVELOPMENT ECONOMICS
Nowotny, K.:
 PUBLIC UTILITY REGULATION
Horowitz, I.:
 ORGANIZATION AND DECISION THEORY
Hennings, K. and Samuels, W.:
 NEOCLASSICAL ECONOMIC THEORY, 1870 to 1930
Samuels, W.:
 ECONOMICS AS DISCOURSE
Lutz, M.:
 SOCIAL ECONOMICS
Weimer, D.:
 POLICY ANALYSIS AND ECONOMICS
Bromley, D. and Segerson, K.:
 THE SOCIAL RESPONSE TO ENVIRONMENTAL RISK
Roberts, B. and Feiner, S.:
 RADICAL ECONOMICS
Mercuro, N.:
 TAKING PROPERTY AND JUST COMPENSATION
de Marchi, N.:
 POST-POPPERIAN METHODOLOGY OF ECONOMICS
Gapinski, J.:
 THE ECONOMICS OF SAVING
Darity, W.:
 LABOR ECONOMICS: PROBLEMS IN ANALYZING LABOR MARKETS
Caldwell, B. and Boehm, S.:
 AUSTRIAN ECONOMICS: TENSIONS AND DIRECTIONS
Tool, Marc R.:
 INSTITUTIONAL ECONOMICS: THEORY, METHOD, POLICY
Babe, Robert E.:
 INFORMATION AND COMMUNICATION IN ECONOMICS
Magnusson, Lars:
 MERCANTILIST ECONOMICS
Garston, Neil:
 BUREAUCRACY: THREE PARADIGMS

Problems of Coordination
in Economic Activity

Edited by
James W. Friedman
University of North Carolina
at Chapel Hill

Kluwer Academic Publishers
Boston/Dordrecht/London

Distributors for North America:
Kluwer Academic Publishers
101 Philip Drive
Assinippi Park
Norwell, Massachusetts 02061 USA

Distributors for all other countries:
Kluwer Academic Publishers Group
Distribution Centre
Post Office Box 322
3300 AH Dordrecht, THE NETHERLANDS

Library of Congress Cataloging-in-Publication Data

Problems of coordination in economic activity /
    edited by James W. Friedman.
      p. cm.—(Recent economic thought series)
    Includes index.
    ISBN 0-7923-9381-3 (alk. paper)
    1. Game theory. 2. Equilibrium (Economics)
I. Friedman, James W. II. Series.
HB144.P76 1993
339.5—dc20   93-14168
                                                                                      CIP

Copyright © 1994 by Kluwer Academic Publishers

All rights reserved. No part of this publication may be reproduced, stored in a retrieval system or transmitted in any form or by any means, mechanical, photo-copying, recording, or otherwise, without the prior written permission of the publisher, Kluwer Academic Publishers, 101 Philip Drive, Assinippi Park, Norwell, Massachusetts 02061.

Printed on acid-free paper.

Printed in the United States of America

Table of Contents

List of Contributors		vii
Preface		ix
Part I	**Introduction**	1
Chapter 1	Introduction and Overview by James W. Friedman	3
Chapter 2	A Review of Refinements, Equilibrium Selection, and Repeated Games by James W. Friedman	17
Chapter 3	Coordination in Games: A Survey Gary Biglaiser	49
Part II	**General Issues in Coordination**	67
Chapter 4	Incorporating Behavioral Assumptions into Game Theory Matthew Rabin	69
Chapter 5	On the Concepts of Strategy and Equilibrium in Discounted Repeated Games William Stanford	89
Chapter 6	The "Folk Theorem" for Repeated Games and Continuous Decision Rules James W. Friedman and Larry Samuelson	103
Chapter 7	Alternative Institutions for Resolving Coordination Problems: Experimental Evidence on Forward Induction and Preplay Communication Russell Cooper, Douglas V. DeJong, Robert Forsythe and Thomas W. Ross	129
Part III	**Coordination in Specific Economic Contexts**	147
Chapter 8	The Dynamics of Bandwagons Joseph Farrell and Carl Shapiro	149

Chapter 9	Dynamic Tariff Games with Imperfect Observability *Andreas Blume and Raymond G. Riezman*	185
Chapter 10	Coordination Theory, The Stag Hunt, and Macroeconomics *John Bryant*	207

Subject index   227

Author index   229

Contributing Authors

Gary Biglaiser, *University of North Carolina at Chapel Hill*
Andreas Blume, *University of Iowa*
John Bryant, *Rice University*
Russell Cooper, *Boston University*
Douglas DeJong, *University of Iowa*
Joseph Farrell, *University of California at Berkeley*
Robert Forsythe, *University of Iowa*
James Friedman, *University of North Carolina at Chapel Hill*
Matthew Rabin, *University of California at Berkeley*
Raymond Riezman, *University of Iowa*
Thomas Ross, *University of British Columbia*
Larry Samuelson, *University of Wisconsin*
Carl Shapiro, *University of California at Berkeley*
William Stanford, *University of Illinois at Chicago*

Preface

A moment's thought will confirm that coordination is extremely important in economic, political, and social life. The concept of economic equilibrium is based on the coordination of producers and consumers in buying and selling, in an orderly way, arrays and quantities of goods that participants expect to buy and sell at prices they correctly plan for. OPEC makes plans from time to time to attempt coordination in the output and pricing of pertroleum. Computer manufacturers must decide whether to adopt technical standards of previously established firms or strike out on their own with different standards. Various examples of coordination have different characteristics and suggest a number of questions. When is coordination achieved and when not? If economic agents coordinate, do they necessarily achieve efficient outcomes? The present volume is concerned with coordination, mostly from a theoretical standpoint.

In the last decade researchers have turned explicit attention to the topic of coordination and it is now clear that there is much work to be done and that great progress in understanding the various facets of coordination is within reach. The aim of the present volume is twofold. First, to contribute to the ongoing research on the economics of coordination and second, to spread results and encourage interest in the topic. Parts II and III of the volume contain original research on coordination. The chapters in part III address general game theoretic questions; that is, questions of general interest across many areas of economics, while the chapters in part III are addressed to particular issues within specific fields of economics (industrial organization, international trade, and macroeconomics). Except for the chapter by Cooper et al. (chapter 7), the chapters are theoretical; the remaining paper is experimental.

Readers will find that the technical level of the chapters varies widely.

Each author has written at the level and in the style that his subject matter required, although all have tried to remain accessible to a general professional audience. All papers are game theoretic or use a game theoretic approach to their subject. While game theory has worked its way into the basic language of economics and is part of the training of a large fraction of doctoral students now, there are still many in the profession whose acquaintance with the subject is relatively slight. In an effort to aid these readers, chapter 2 provides a review of most of the game theoretic concepts and results upon which later chapters draw. As the chapters in parts II and III are individual efforts by their various authors, the chapters in part I are designed to provide the background which the nonspecialist would typically need to read the later parts with profit. To that end, chapter 1 is an overview of coordination and roadmap to the other chapters. Chapter 2, as noted, is game theory, and chapter 3 is a general background chapter on coordination that explains the character of coordination and reviews important previous results.

Apart from the thanks I give to various people for comments on chapter 2, I would like to thank Warren Samuels for suggesting this endeavor in the first place and Zachary Rolnik for patience, understanding, and encouragement while it has been carried out. Above all, I owe a large debt to all the contributors. They have written fine, interesting, original, and thoughtful papers for this book. Obviously without their efforts, there is simply nothing to present. This book is, itself, a strongly positive act of coordination; all those who agreed to contribute to it have, in fact, done so. And the time frame within which they have operated has not been very long.

I INTRODUCTION

1 INTRODUCTION AND OVERVIEW

James W. Friedman

The *Oxford English Dictionary* defines *coordinate* as: "To act in combined order for the production of a particular result." The discipline of economics could be viewed as analyzing the coordination of economic life and economic equilibria are often portrayed as harmonious outcomes. Obvious examples are equilibrium in a single market (when *the amount supplied* equals *the amount demanded*) and the general competitive equilibrium of an economy à la Walras (1926/1954). Indeed Walrasian general equilibrium is an explication of Smith's (1776/1976) *invisible hand*. Smith (1776/1976, p. 145) provides another example of coordination in his allegation of a tendency of businessmen in any specific market attempting to collude when they have the chance.

Economics focuses principally on the outcomes that may result from coordination, but has little to say concerning the means by which coordination is achieved or which of several candidates for a coordinated outcome will be observed. The (static) stability analysis in general equilibrium, stemming from both the Walrasian tâtonnement and the stability analysis of Cournot (1838/1980) (see Arrow and Hahn 1971, chs. 11, 12), is an early attempt to get at the following question: On which equilibrium do we coordinate, and how do we achieve it? The papers in the present volume are largely concerned with the latter question in both a general game theoretic framework and within specific areas of economics.

Games can be classified into three groups according to the inherent conflict of interest among the players. At one extreme are *two-person strictly competitive games*. These are games in which every possible outcome is Pareto optimal and there is no possibility of two or more players making common cause against another player. Comparing two outcomes A and B in such a game, if A is preferred to B by player 1, then the reverse holds for player 2. There is absolutely no common ground among the players; hence, no room for coordination. Two person zero-sum games are the classic examples.

At the other extreme are *games of common interests* in which all players rank outcomes in precisely the same fashion. Thus, if player 1 prefers A to B, then all players prefer A to B. Such games, surprisingly, might provide coordination problems despite the commonality of outlook. For example, the players might have difficulty in carrying out their common interest if communication between them is limited and there are different ways that they might achieve their most favored outcome.

In between these extremes stand *games of mixed interests*, which include most of the games that naturally arise in economics. In these games, the players' interests partly coincide and partly conflict. A prominent example is a competitive general equilibrium model of pure trade. All players share an interest in achieving a Pareto optimal allocation of goods, but interests are surely in conflict among the Pareto optimal allocations. Similarly, in bilateral bargaining over a single, indivisible object, when the seller's reservation price is below the buyer's, they have a mutual interest in making a trade, but they have opposing interests concerning the price at which trade occurs. Note that even a three-person zero-sum game provides the opportunity for two players to have a common interest in gaining at the expense of the third; therefore, such games fall into the large category just discussed.

In the remainder of this chapter, section 1 is devoted to a brief look at several species of coordination problem and section 2 gives a sketch of the later chapters.

1. Examples of Coordination

In everyday life we see numerous ways in which coordination is effected, along with numerous ways that coordination fails. Successful examples include a) the custom of driving vehicles on one particular side of roads (e.g., on the right in most places) and b) customary business hours, which vary somewhat from one country to another, but which are quite consistent within single countries. In both cases, it may not be important which of several conventions emerges; it only matters that some particular convention does emerge. The British drive on the left and the French drive on the right. Both are quite workable, and given that you are in (say) England, it is in your own interest to drive on the left. Note, however, that trouble could ensue if you were going to be driving and had no way of telling which country you were in. The cost of not knowing where you are could be that you have an accident. Common business hours for office workers such as lawyers, accountants, business offices of commercial companies, and so on, facilitate communication and reduce the cost of operation.

It is true that driving is regulated by law, giving rise to several pertinent observations. First, coordination may sometimes be achieved through legislation. Second, legislation might sometimes give legal sanction to forms of coordination that have arisen spontaneously. Third, to see that coordination can arise without the intervention of legal force, note that pedestrians tend to pass one another on the right in the United States and continental Europe and on the left in England. Law is not binding for pedestrians, but the driving customs suggest a natural coordination for walkers. Similarly, in large, open parking areas drivers tend to pass one another as on streets even though law does not designate a rule for them.

In the preceding examples an individual cannot be acting in her own interest if she chooses a different action than the other agents. A driver in England will be worse off driving on the right than driving on the left. Keeping office hours from, say, midnight to eight in the morning will be worse for a firm than keeping the same hours as other firms. These are games of common interest because there is no conflict of interest among the individuals and equilibrium behavior necessarily implies globally Pareto optimal behavior. When coordination problems arise in such games, they are called *pure coordination problems* due to the coincidence of players' interests. That pure coordination problems can easily arise is illustrated by a well-known example involving two people who have arranged to meet in New York City at 3 P.M. on a particular day, who have neglected to name a meeting place, and who have no means of communicating. Suppose they each achieve a utility of one if they meet and zero if they fail to meet. Where should each appear at 3 P.M. on the appointed day? All equilibria are equally good and any equilibrium is characterized by the two people choosing the same place. Suppose there are two obvious places to meet: on the steps of the Metropolitan Museum and on the steps of the Public

Table 1–1. A Pure Coordination Problem

		#2 museum	#2 library
#1	museum	1 1	0 0
	library	0 0	1 1

Library. This situation is illustrated in table 1–1 where agent 1 chooses the row and agent 2, the column. In each cell, the first utility is for agent 1 and the second for agent 2. These two players want to do the same thing (i.e., to get to the same meeting point), but they require a coordinating mechanism to get them to reliably make the same decision. There is no efficient answer to the problem which is posed here as a one-shot occurrence. Arguably each player will be equally likely to select either alternative, causing the probability that they meet on time to be 0.5.

The preceding situation becomes different if the circumstances will be repeated many times; therefore, now suppose that these people are to meet each Monday afternoon at 3 P.M. but they cannot ever discuss where to meet (e.g., they have no common language). Under what circumstances will they eventually settle on a convention (i.e., on a meeting place)? Intuition suggests that they will do so after a moderate number of trials; however, just how they do this is not clear. If and how players can coordinate when they do not have a common description of the game that they can discuss, even when they will play the game repeatedly and their interests are absolutely identical, is not a trivial question. Crawford and Haller (1990) deal with this issue in an interesting paper that is discussed in chapter 3 of the present volume.

The driving example can be cast in a somewhat different way. Imagine that there are many drivers and, from time to time, pairs of them encounter one another. When two drivers encounter one another, each coming from a different direction on a road, they either pass safely or crash. In the former situation they both choose *right* or they both choose *left*, but in the latter situation, one chooses *left* and the other *right*. Even if the two people both choose right, the general problem of societal coordination is not solved. Furthermore, that these two people continue to choose right in the future also does not solve the general problem. After all, other pairs of drivers may have successfully avoided crashes by choosing left. Again, a question arises concerning the way that coordination could, or must, develop. Somehow it is necessary that *all drivers* adhere to the same

Table 1–2. The Battle of the Sexes

		W ballgame		concert	
H	ballgame	2	1	0	0
	concert	0	0	1	2

passing standard. Intuition suggests that repeated encounters will lead to all drivers adopting the norm that is followed by the population majority if each person revises his own choice periodically to conform with the majority of drivers she has recently encountered. But whether this conjecture is true is not clear and its truth may depend upon the majority being sufficiently large or on the frequency with which a person revises her choice being sufficiently great (or small) or on the drivers adjusting slowly enough to avoid unstable oscillations in the fraction of drivers choosing right. This simple driving problem is not trivial and is representative of many social situations.

A second sort of coordination problem is exemplified by a famous game called the *battle of the sexes*, shown in table 1–2. The game assumes stereotyped preferences for a married couple who must decide on their Saturday night entertainment. There are three possibilities: go to a ballgame together, go to a concert together, and stay home together. The husband's (H) preferences order these, from best to worst, ballgame, concert, home, and the wife (W) orders them concert, ballgame, home. It is assumed that they would both prefer to stay home together rather than to go alone to either of the other activities and that, if they fail to agree, they will stay home. Of course, real people often solve this problem by going occasionally to ballgames and occasionally to concerts. This is not considered here; the game is strictly one-shot; furthermore, discussion to reach consensus is also ruled out. Instead, they choose simultaneously and each player is to choose either *ballgame* or *concert* with the understanding that they stay home if they choose differently and they go to the chosen activity if they both select the same thing. Assigning utilities of 2, 1, and 0 respectively to the three alternatives, it is clear that there are two Pareto optimal equilibria. They are 1) both choose *ballgame* and 2) both choose *concert*. As they must choose simultaneously, each chooses without knowing for certain what the other will select.

The heart of the matter here is that they may easily choose differently, resulting in an outcome that is neither an equilibrium nor is Pareto optimal.

Table 1-3. The Cuban Missile Crisis

		USSR turn back		USSR sail on	
US	blocade	2	1	−1	−1
	not block	0	0	1	2

Table 1-4. Prisoners' Dilemma

		#2 confess		#2 not confess	
#1	confess	2	2	9	0
	not confess	0	9	5	5

Although the two *pure strategy* equilibria (both choose *ballgame* and both choose *concert*) are Pareto optimal, neither of the two equilibria stands out over the other as the obvious joint choice. Although there is a third equilibrium under which each player randomizes, this equilibrium does not guarantee coordination. It would be possible to achieve coordination if they could randomize together (e.g., heads they go to the ballgame, tails to the concert). Such a possibility is reasonable in the particular circumstances sketched here; however, it is easy to visualize situations in which the payoff structure of table 1-2 prevails, but the participants cannot be expected to randomize jointly. Indeed, some games involving confrontation have nearly the same formal structure as the battle of the sexes. The celebrated Cuban missile crisis of the early 1960s could be seen this way. The United States can choose between blockading Cuba and not blockading, while the USSR chooses between sailing to Cuba and not. This is shown in table 1-3. In this game, there are two ways the players may fail to coordinate, each with different payoffs; however, the game is strategically very much the same as the battle of the sexes. It is difficult to imagine Kennedy and Khrushchev agreeing on the telephone to flip a coin with heads leading to one equilibrium and tails to the other!

A third circumstance in which interesting coordination problems arise is illustrated by the *Prisoner's dilemma game*, shown in table 1-4. Whatever agent 2 might choose, agent I will be better off choosing *confess*; however, the same holds for agent 2. Thus the game has a unique equilibrium at which the players both confess and receive payoffs of two each. However, the equilibrium utilities (2, 2) are inefficient, because (5, 5) is obtainable. In this situation both players recognize they will both be better off choosing *not confess*, but they each have individual incentives that push them toward *confess*. The question here is whether something could alter or augment their incentive structure to get them to behave in a more jointly beneficial manner. As long as this game is played just once, there appears to be no way to avoid the inefficient outcome under which both players confess. If the game is to be played repeatedly, more possibilities open up and equilibria can be found that achieve the efficient outcome.

The key feature of the prisoners' dilemma is that the equilibrium is inside the payoff possibility frontier. Many single shot games share the feature that all Nash equilibria are inefficient. It is difficult to see how any way around this inefficiency could be found as long as the game is single shot. That agents behave in their own self-interest is very compelling, and it is difficult to see how the self-interest of a player in, say, the prisoner's dilemma is served by not confessing. Even though both are better off when no one confesses as compared with both confessing, the incentives of the individual compel confession. Repetition permits coordination on Pareto optimal outcomes by allowing a fundamental change in the incentive structure of the game. The players can follow strategies that call for a Pareto optimal choice in each iteration and provide for punishment if a player deviates. Although coordination becomes possible with repetition, many games have an abundance of Pareto optimal outcomes, injecting a coordination problem somewhat like that of the *battle of the sexes*.

In summary, at least three sorts of coordination problem can be delineated. The most uncluttered is the *pure coordination problem*, arising in games of common interests, in which there are efficient equilibria, all of which are payoff equivalent. In these games, players are all indifferent about which efficient equilibrium they reach and are happy to reach any one of them. Their problem is to simply reach one of them. Table 1–1 provides an example. These coordination problems are easily solved if the players have a common understanding of the description of the game and they can also talk (engage in preplay communication) before selecting strategies. Thus, a problem arises when they cannot talk or cannot speak a mutually understood tongue. The latter would hold if the two players in table 1–1 could talk, but had no words for *library* and *museum*.

The second coordination problem arises in games of mixed interests and is characterized by having several Pareto optimal equilibria that are ranked differently by the players. This might be dubbed a *competing preferences coordination problem*. It is exemplified by tables 1–2 and 1–3.

Table 1-5. Prisoners' Non-Dilemma

		#2 confess		not confess	
#1	confess	2	2	3	0
	not confess	0	3	5	5

Even if the players can talk with one another, there is no assurance they can come to an agreement on an equilibrium. At least, if they could agree on a specific equilibrium, one could expect them to adhere to their agreement.

The third problem is that of *coordination to avoid inefficiency*, arising in games of mixed interests and exemplified by the *prisoners' dilemma* shown in table 1-4. In discussing this above, I took the position that coordination on the efficient outcome was impossible as long as the game was to be played just one. There are two objections to this position that I want to raise and comment on. The first is the objection based on *public spirit* under which it is argued that the players might sufficiently value efficiency and the general good to overcome their selfish incentives. (If we all cease tossing our empty beer cans on the roadside, we will not have to pay for cleaning crews.) If the players possessed such public spirit, then the payoffs shown in table 1-4 would not reflect the players' true utilities. Suppose that each player would feel so badly about being the one to ruin coordination on *not confess* that she would rather receive the *not confess/ not confess* outcome than the outcome she achieves when she confesses and the other player does not confess. Were this the case, her payoff when she alone confesses must be lower than her payoff when neither confesses. The payoffs might then look something like those of table 1-5. Table 1-5 has two equilibria: The original *confess/confess* equilibrium and *not confess/not confess*. The point, of course, is not to say that being public spirited is either silly or irrational or impossible. The point is that the degree and form of public spirit affects the utilities associated with different outcomes and the payoffs to the game are the actual utilities experienced by the players. Thus, the presence of public spirit can mean the payoffs are similar to those of table 1-5. In such a case, the game is no longer the prisoners' dilemma and the problems associated with the prisoners' dilemma are absent.

The second objection is that, though the prisoners' dilemma may be played on a single-shot basis, the players remain members of a larger community in which they interact from time to time with one another and with other members of the community. Consequently, they might cooperate in not confessing because they will be in other games in the future and their reputations, gained from their behavior in earlier games, will affect the expectations of the players with whom they are engaged. As with community spirit, this objection raises something real, something that game theory and economics should strive to understand. Economic agents are in a variety of situations from time to time, or are even in several simultaneous situations. But a game should be analyzed as a stand-alone entity; therefore, if possible future interactions are relevant to strategy choice in a (supposedly) one shot prisoners' dilemma, then that one shot prisoners' dilemma is not the relevant game to be analyzed. It is merely a piece of a larger game, and that larger game should be the object of study. In brief, the real answer to the objection is to develop the ability to analyze more complex games.

2. Overview of the Book

This volume neither addresses nor solves all possible coordination problems; however, it does present current, original research on coordination. Part I of the book, comprising chapters 1–3, provides useful background for and an introduction to parts II and II. Chapter 2 is a review of some aspects of game theory that should aid the reader in recalling concepts and results that are used in later chapters. It can be used for occasional reference or for a more concentrated general review, according to the reader's needs. Chapter 3, by Gary Biglaiser, gives a concise description of some interesting and important papers in the coordination literature. These deal with such topics as pure coordination, conditions under which equilibrium in a repeated game must involve the achievement of coordination, and the exploration of some laboratory experiments on coordination.

Part II comprises four chapters concerned with coordination from a general game-theoretic perspective, while part III consists of three chapters that deal with specific economic contexts. Chapter 4 by Matthew Rabin begins part II and develops theoretical tools permitting behavioral assumptions and beliefs to be merged with usual considerations of rationality in an effort to better model the manner in which rational individuals make decisions. Rabin constructs what he calls a *consistent behavioral theory* in which players may, for example, confine their behavior and their beliefs about the behavior of the other players upon strategies leading to Pareto optimal outcomes. The general point is that there are subsets of the players' strategy sets to which attention can be confined. These subsets are not unique; however, they are not arbitrary. The subset of strategies to which players believe player *i* will restrict himself must include all strategies that are rational in the light of player *i*'s presumed beliefs about the choices that other players might make. And, the beliefs of all the players must be consistent with one another.

In chapter 5 William Stanford investigates two common definitions that are given to the concept *strategy*. He looks at the connection between these definitions and subgame perfect Nash equilibrium in repeated games to see whether our notions of rational play imply that one of these definitions is more appropriate. The more comprehensive definition requires that a strategy indicate what move a player will select at every decision point (i.e., information set) in the game where that player moves. This can include specifying moves at decision points well into the game that are impossible to reach given the earlier moves by that same player. For example, suppose player 1 must choose heads (H) or tails (T) early in the game and then, if H was chosen, will eventually arrive at a point where the same player must choose between vanilla (V) and strawberry (S). The most comprehensive definition of strategy requires that the strategy indicate a choice between V and S even when the strategy calls for T at the earlier point, making it impossible that the V/S choice will even be operational. A less demanding notion of strategy only requires decisions to be specified for contingencies that can actually occur.

Stanford argues that it is important to use the more comprehensive definition. The intuition behind his reasoning runs along these lines: What is best for player 1 above depends on the strategy that player 2 will follow; however, the strategy that player 2 finds best can depend upon what player 2 believes player 1 would do if player 1 reached the V/S choice. We want the V/S choice to be part of player 1's strategy so that the two players' strategies from the V/S choice onward consist of best replies to one another. Then, the choice of player 1 to select T at the start is part of a best reply for player 1; that is, to choose H and eventually reach the V/S choice will, in the end, be worse for player 1 than to have chosen T. Such a comparison cannot be discussed if the strategies of the players do not specify moves at all decision points.

Chapter 6, by Larry Samuelson and me, proposes particular forms of repeated game strategies as being especially reasonable forms by which players might achieve coordination on Pareto optimal outcomes in generalized versions of the repeated prisoners' dilemma. The ability to achieve Pareto optimal outcomes in such games is one of the important implications of the *folk theorem for repeated games* and the extentions of that result; however, the theorems in this area rely upon strategies that involve discontinuous decision rules. Typically, coordination requires a precise collection of choices by the players and any deviation from the precisely expected choices leads to a dramatic change in the players' behavior. The smallest deviation by a player is met with the same response as the largest possible deviation. While these strategies are logically and technically unassailable, they are not intuitively convincing. That is, the typical equilibrium strategy combination calls for players to coordinate upon a particular joint action. Any deviation from this joint action, no matter how small or large, by a player is met with a severe punishment that is chosen to be large enough to make any such deviation unprofitable. Friedman and Samuelson argue that it is implausible for players to engage in precisely the same punishment irrespective of the size of the deviation; for this means that tiny deviations will be punished with huge penalties. This is like executing a person for overtime parking.

Friedman and Samuelson are able to prove theorems like those in the *folk theorem* literature, but that are based upon continuous decision rules under which small deviations by a player are met with small punishments. Thus the size of the punishment is continuouosly related to the size of the deviation. As the size of the deviation goes to zero, so also does the punishment. This maintains an intuitively reasonable proportion between the two.

In chapter 7, the final chapter in part II, Russell Cooper, Douglas DeJong, Robert Forsythe, and Thomas Ross report on laboratory experiments in which they investigate the roles of preplay communication and forward induction in determining the behavior of subjects. Forward induction refers to the inferences a player makes about the likely future behavior of his rivals, based upon the behavior that has already been observed. In games that are variants on the battle of the sexes forward induction sometimes appears to provide convincing evidence of what a player will do. With respect to preplay communication, it is vital to remember that it is, by definition, nonbinding; therefore, it is natural to question whether the statements of a player should be believed. Cooper et al. are concerned to determine the conditions under which forward induction is an aid to coordination and also to understand how forward induction affects the choice of equilibrium. Similarly, they seek to determine the conditions under which preplay communication affects the choice of equilibrium and how it does so.

The papers of part III are each directed at a specific economic problem. Chapter 8 by Joseph Farrell and Carl Shapiro focuses on a timely problem in industrial organization. Suppose two or more firms innovate in the same area, but do so with somewhat different technological standards so that the product of one firm is incompatible with the product of others. Videocassette formats, compact disc formats, computer processors and operating systems, and many other products are current examples. Coordination on a single standard will cause the total market to be larger by making the products more useful to consumers, but each firm would be better off if its own particular standard were the industry standard. How, when, and in what manner does coordination take place? This is roughly like an intertemporal variant of the battle of the sexes. If or when a firm should abandon its own standard and adopt that of a rival depends upon both the market potential of each technology and the previously installed base of each.

Chapter 9 by Andreas Blume and Raymond Riezman takes up tariff policy in a repeated game context somewhat like that of chapter 6 with a vital difference. The trade policies of the countries are only imperfectly observable. In such a setting can the inefficiencies inherent in the prisoners' dilemma be avoided? Or at least partially avoided? The *folk theorem for repeated games* and related results show how efficient equilibria can be attained when there is perfect monitoring (i.e., when the trade policies of the countries can be perfectly observed); however, Blume and Riezman are dealing with a model in which the previous period's actions by the other country are not directly seen. Only an imperfect signal is seen. They show reasonable conditions under which Pareto optimality cannot be attained, although outcomes superior to single shot equilibrium outcomes are attainable.

The final chapter, by John Bryant discusses macroeconomic policy making using a game due to Jean-Jacques Rousseau as a vehicle. The game is one in which each of several players must decide on an effort level with total production being determined by the smallest effort level chosen. That is, the effort of each player can be regarded as a separate input to production with production requiring fixed and equal proportions of the several inputs. Any strategy combinations under which all players choose identical effort levels is an equilibrium, because effort is costly to a player and, when a player's effort level is higher than that of some other player, the marginal product of his effort is zero. At the same time, the payoff to each player in equilibrium increases as the equilibrium effort level increases. There is a best equilibrium at which all players expend the maximum effort level. This equilibrium Pareto dominates all other equilibria; however, with many players involved it may be difficult to achieve coordination among the players at this equilibrium.

References

Arrow, Kenneth J., and F.H. Hahn. 1971. *General Competitive Analysis*. San Francisco: Holden-Day.

Cournot, A.A. 1838/1980. *Recherches sur les Principes Mathématiques de la Théorie des Richesses*. Gérard Jorland (ed.). Paris: J. Vrin.

Crawford, Vincent P., and Hans Haller. 1990. "Learning How to Cooperate: Optimal Play in Repeated Coordination Games," *Econometrica* 58: 571–595.

Smith, Adam. 1776/1976. *An Inquiry into the Nature and Causes of the Wealth of Nations*. R.H. Campbell, A.S. Skinner, and W.B. Todd (eds.). Oxford: Clarendon.

Walras, Léon. 1926/1954. *Elements of Pure Economics*. Translated by William Jaffé. Homewood: Irwin.

2 A REVIEW OF REFINEMENTS, EQUILIBRIUM SELECTION, AND REPEATED GAMES*

James W. Friedman

1. Introduction

The purpose of this chapter is to provide both a review for the relatively well-prepared reader and an introduction to the parts of game theory used in subsequent chapters for the reader who has relatively little game theory background. The exposition assumes readers who have a rudimentary background; that is, who know what constitutes a game, know the meaning of *move* and of *strategy* and how they are different, and so forth. In the spirit of a review of concepts and results, nothing is proved and, in many instances, models and concepts are explained intuitively and without technical completeness. The reader who needs and wants more is urged to read the very good available textbooks and original sources.

---

* This chapter has benefitted from the comments of Dennis Coates, Matthew Rabin, Ray Riezman, Tom Ross, Larry Samuelson, and Bill Stanford. The remaining shortcomings are my responsibility.

1.1. Chapter Overview

In section 2 the basics of finite extensive form games and of strategic form games are reviewed. For this material other sources include the textbooks of Binmore (1992), Friedman (1990), Fudenberg and Tirole (1991), Luce and Raiffa (1957), and Myerson (1991). Section 3 takes up three equilibrium refinements, *subgame perfection, perfection, and sequential equilibrium*. The reader wishing more material on these and other refinements will find help in all the preceding books except Luce and Raiffa, which was published before refinements were born. The best single source, however, is probably van Damme's (1987) fine treatment, which covers many refinements. And there are original sources such as Selten (1965) for subgame perfection, Selten (1975) for perfection, and Kreps and Wilson (1982) for sequential equilibrium.

Much of section 4 is concerned with very recently developing material including *forward induction*, which was put into active play by Kohlberg and Mertens (1986), *cheap talk*, which leapt into deserved importance from the work of Farrell (1987, 1988), and *rationalizable strategies*, which was co-discovered independently by Bernheim (1984) and Pearce (1984). Repeated games and the results linked to the phrase "folk theorem for repeated games" is sketched in section 5. Some of the recent texts cover this material. The central articles on which section 5 draws are Abreu (1988), Friedman (1971), and Fudenberg and Maskin (1986); however, other historically important original sources include Aumann and Shapley (1976) and Rubinstein (1979).

1.2. Relationship to Later Chapters

Section 2 is a general introduction to basics in game theory that is relevant to all later chapters. Section 3 on refinements is relevant to chapters 5, 6, 8, and 9, particularly regarding subgame perfection. Section 4, which explores alternatives to the standard refinements, is relevant to chapters 4, 7, and 8. Section 5, on repeated games, is relevant to chapters 5, 6, 8, and 9.

2. Finite Games in Extensive and Strategic Forms

The *extensive form* and the *strategic form* are two common ways to represent games. The extensive form is a detailed representation in which each individual move, or act, in which any player might engage is explicitly represented. Each point in a game at which a player can take a specific action is called a *decision node*. Points at which the game ends are called *terminal nodes*. In the strategic form the move by move detail is not evident; instead, the players' choices and possible actions are condensed into *strategies*. A strategy of a player is a comprehensive plan indicating the action to be taken by the player in each of the specific decision circumstances that can arise for the player in the game.

A game is *finite* if it has a finite number of players and there are a finite number of moves that any player might make. A common example is tic-tac-toe, in which the first player has nine moves to choose from when the game begins. Then, after the first move of player 1, player 2 could find herself at any one of nine positions (*decision nodes*). At each of these positions, player 2 has eight possible move. Following the first move of player 2, player 1 could find himself at any of 72 possible positions, depending on which of his first nine possible moves he chose and which of player 2's possible eight, from which he has seven possible moves. And so forth. A game is finite if the total number of positions (decision nodes) in the game is finite and the number of choices at each of them is finite. In this section, extensive form games are described first, then strategic form games will be developed from the extensive form.

2.1. Games in Extensive Form

The extensive form of a game is characterized by several entities: a) the *game tree*, Y, which shows the decision structure of the game; b) the set of players, $N = \{1, \cdots, n\}$; c) the payoffs, p, to each player associated with each endpoint (*terminal node*) of the game; d) the information, I, the players will have at each decision node concerning which moves have been chosen thus far; e) the rules governing random occurrences, q, in the game; and f) the information that players have about the game apart from item d above.

The game tree, Y, is characterized by the set D of decision nodes in the game, the decision node d_0 at which the game begins, the set T of terminal nodes at which the game ends, and the rules governing movement from one node to another. Basically, in a finite game these rules determine the class of possible game trees: i) D and T have a finite number of elements, and ii) from d_0 to any other node $d \in D \cup T$ there is a unique path. Figure 2–1 provides an example. The cross-hatched circle at the bottom is d_0, the other open circles are the other decision nodes, the filled circles are the terminal nodes, and the straight lines between nodes indicate moves. The

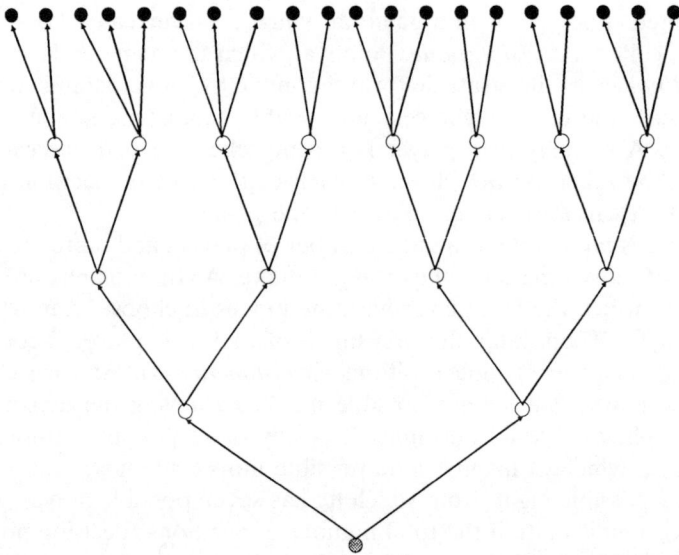

Figure 2–1. A Game Tree

arrowhead at the end of each straight line indicates the direction of movement. The payoff function, p, associates a payoff vector $p(t) = (p_1(t), \cdots, p_n(t)) \in \mathbf{R}^n$ with each terminal node $t \in T$. In a game, each decision node $d \in D$ is associated with the particular player who moves at that node. An important special class of games are *games of perfect information* in which each move made by a player is observed by all other players when it is made and is remembered by all players. Figure 2–2 is an example. Each decision node in the figure is labelled with the player who moves at that position, the possible moves are indicated by arrows leading away from the node, and each terminal node has a payoff vector associated with it. *Perfect information* means that, at each node $d \in D$, the player who moves there knows that he is at that particular node. *Imperfect information* will be illustrated below. This game has no randomness in its structure, so e above will be temporarily ignored. As to item f, most of this chapter will deal with games of *complete information* in which complete information is assumed to be *common knowledge*. Complete information means that every player knows the structure and payoffs of the game. That is, in a game such as the one in figure 2–2, each player has a copy of the figure (or has equivalent information). Thus each player knows who the players are, who

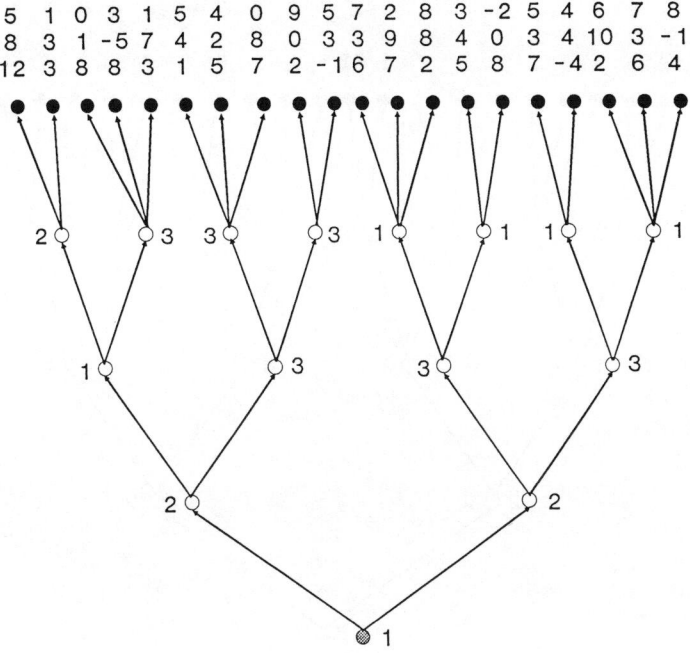

Figure 2-2. A Game of Perfect Information

moves at each node, what moves are available, and what payoff vector obtains at each terminal node.

Common knowledge will be explained here in a usual, intuitive fashion. If all players were brought together in a room and, in full view of one another, each was handed a copy of figure 2–2, then complete information would be common knowledge. Contrast this with the following situation. Each player is placed alone in a room that has a one way window giving view into a central room. One by one, each player is brought from her room to the central room where she is given a copy of figure 2–2 in full view of the remaining players and is then brought back to her own room. Each player would have complete information, each would know that the other players had complete information, but no player would know if the other players knew whether they all had complete information. Even if she were told the other players knew she had complete information, she would not know whether they knew that she knew that they knew that she had complete information. When complete information (or some other matter

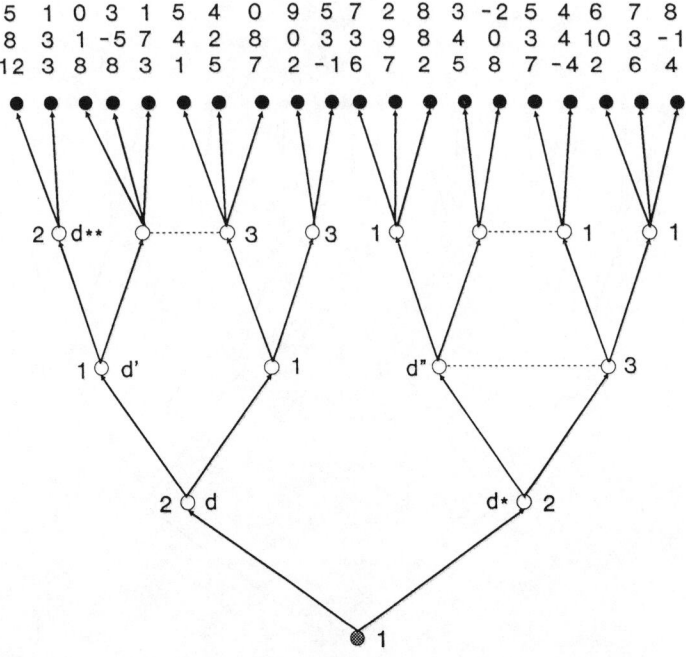

Figure 2–3. A Game of Imperfect Information

of information) is common knowledge, then everyone knows, everyone knows that everyone knows, everyone knows that everyone knows that everyone knows, and so forth.

Sometimes certain moves of players are not observed by one or another of the players. In matching pennies, for example, the two players choose their moves simultaneously and the extensive form representation is made on the pretense that one player moves first, but is unobserved by the other. Figure 2–3 provides an example. Notice that certain decision nodes, belonging to only one player, are linked together by broken lines. This indicates that the nodes thus linked are indistinguishable by the player who moves there. That is, the player knows too little about past moves to be able to tell which of these nodes is the actual one where he is located. If several nodes are indistinguishable in this way, then the number of moves at each of these nodes (and the physical meaning of the moves) must be identical. Such a grouping of nodes is called an *information set*. In a game of perfect information each information set contains exactly one node. In a game of imperfect information there must be at least one information set containing two or more nodes. The collection of information sets is denoted I and the elements of I comprise a partition of D, the set of decision nodes. It is sometimes helpful to think of grouping the elements of I by player. Thus we may denote the family of information sets of player i by I^i and write $I = \{I^1, \cdots, I^n\}$. Then $I^i = \{I_{i1}, \cdots, I_{ir_i}\}$ where each I_{ij} is a specific information set of player i.

2.2. Deriving the Strategic Form from the Extensive

If a game has no randomness in its structure, then it can be described as $\Gamma = (N, Y, p, I)$. To give the *strategic form* description of such a game requires specification of the sets of *pure strategies* of the players, the sets of *mixed strategies* of the players, and the *payoff function* of each player, which expresses the payoff of the player as a function of the strategies chosen by all players. A *pure strategy* is a strategy that involves no voluntary randomization by the player; therefore, at each information set in the game where player i makes a move, a pure strategy selects a specific move. Thus, in figure 2–2, player 1 has 6 information sets, player 2 has 3, and player 3 has 6. In figure 2–3, player 1 has 6, player 2 has 3, and player 3 has 3. The number of distinct pure strategies for player 1 in, for example, the game of figure 2–3, is 144. This is because she has two moves in each of three information sets and three moves in the other two information sets, giving her $2 \times 2 \times 2 \times 2 \times 3 \times 3 = 144$ pure strategies. Likewise, player 2 has $2 \times 2 \times 2 = 8$ pure strategies and player 3 has $2 \times 2 \times 3 = 12$.

The set of pure strategies of player i can be denoted $\Pi_i = (1, \cdots, m_i)$ where m_i is the number of pure strategies of player i. Thus each integer in Π_i is identified with a distinct pure strategy. $\Pi \equiv \times_{i \in N} \Pi_i$ is the pure strategy space of the game, Let $a^\pi = (a_1^\pi, \cdots, a_n^\pi) \in \mathbf{R}^n$ denote the payoff vector associated with $\pi \in \Pi$. The set S_i of mixed strategies of player i consists of the set of all possible probability distributions over the elements of Π_i. A typical element of S_i is $s_i = (s_{i1}, \cdots, s_{im_i})$ where s_{ij} is the probability that player i places on her jth pure strategy. Therefore, $S_i = \{s_i \in \mathbf{R}_+^{m_i} | \sum_{j=1}^{m_i} s_{ij} = 1\}$ and $S \equiv \times_{i \in N} S_i$. The payoff function of player i is

$$P_i(s) = \sum_{\pi_1=1}^{m_1} \cdots \sum_{\pi_n=1}^{m_n} s_{1,\pi_1} \cdots s_{n,\pi_n} a_i^\pi \tag{1}$$

Letting $P = (P_1, \cdots, P_n)$, a game in strategic form is given by $\Gamma = (N, S, P)$.

Structural uncertainty is introduced into the game through the medium of an artificial player known as either *player 0* or *nature*. The formal incorporation of nature into the extensive form of the game is by treating nature just like any other player except in these ways: a) nature receives no payoffs, b) there can be only one decision node in each information set at which nature moves, and c) at each of nature's information sets, nature chooses according to a probability distribution that is part of the description of the game. Thus, suppose there are r_0 information sets at which nature moves. At the first of these, the choice of move is determined by the probability distribution $q^1 = (q^1_1, \cdots, q^1_{m_{01}})$ where m_{01} is the number of moves at this information set. Of course $q^1 \geq 0$ and $\sum_{i=1}^{m_{01}} q^1_i = 1$. In general, at the kth information set of nature, the choice of move is determined by the probability distribution $q^k = (q^k_1, \cdots, q^k_{m_{0k}})$ where m_{0k} is the number of moves at this information set, $q^k \geq 0$ and $\sum_{i=1}^{m_{0k}} q^k_i = 1$. Letting $q = (q^1, \cdots, q^{r_0})$, the extensive form game is given by $\Gamma = (N, Y, p, I, q)$. The payoff functions for the strategic form are still given by equation q; however, the a_i^π are the expected payoffs associated with π, taking the expectation over the moves of nature. Figure 2–4 provides an illustration. There are two information sets at which nature moves, labeled with "0." The moves from these information sets are labeled with the associated probabilities.

A last definition in the basic description of extensive form games is *perfect recall*. In a game of perfect recall, a player never forgets any move he made and he does not forget anything that he learned about previous moves made by the other players. Both figures 2–2 and 2–3 show games of perfect recall; however, figure 2–4 shows a game of imperfect recall. If player 1 reaches his information set containing three nodes, he will have forgotten whether he chose right or left at the beginning of the game. Note that, immediately following the first move of player 1, player 2 will be at either of two information sets. She will know whether player 1 chose right or left at the start of the game. But if player 2 reaches the information set containing six nodes, she will have forgotten what play 1 chose at the start. She will also have forgotten what she chose following the first move of player 1. Our concern in this chapter will be exclusively with games of perfect recall.

In addition to choosing a mixed strategy of the kind defined above, a player can choose to randomize in a different fashion; the player can choose a *behavior strategy*. Such a strategy consists of several independent probability distributions, one for each of the player's information sets. Denote by r_i the number of information sets at which player i moves and by m_{ij} the number of moves in the jth information set of player i. Let $\beta^i_j = (\beta^i_{j,1}, \beta^i_{j,2}, \cdots, \beta^i_{j,m_{ij}})$ be a probability distribution over the moves in the jth

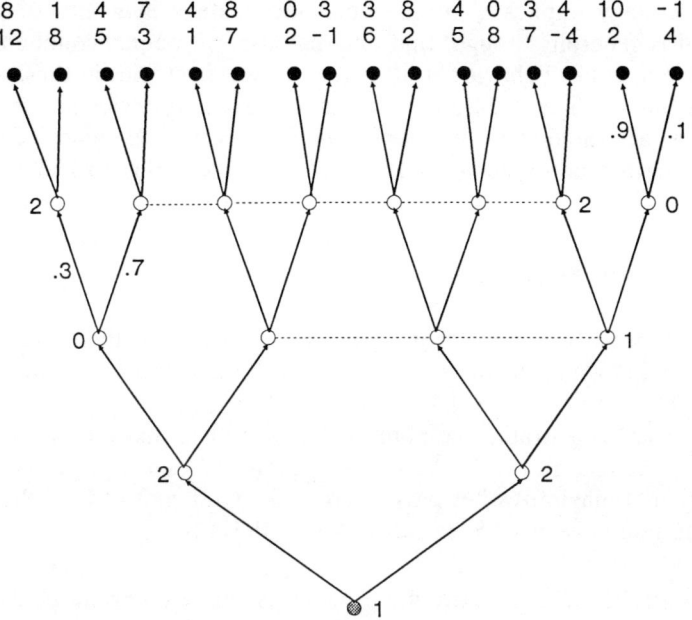

Figure 2-4. A Game of Imperfect Recall

information set of player i. Thus $\beta_j^i \geq 0$ and the elements of β_j^i sum to one. Then $\beta^i = (\beta_1^i, \beta_2^i, \cdots, \beta_{r_i}^i)$ is a *behavior strategy for player i* and $\beta = (\beta^1, \beta^2, \cdots, \beta^n)$ is a *behavior strategy combination*. An important question is whether there is any sort of equivalence between mixed strategies and behavior strategies. In games of perfect recall, there is a close relationship, though one which falls short of equivalence. In such a game, a mixed strategy implies a unique behavior strategy for the player. That is, a mixed strategy of player i implies a unique probability distribution over the moves at any of player i's information sets, conditional upon reaching that information set. It is possible that several different mixed strategies will induce (imply) the same behavior strategy. In a game of *imperfect recall*, the implied behavior strategy of a player can depend upon the strategies of other players, as well as upon her own strategy.

Intuitively, the ability to alter one's chosen course of action as the game progresses would appear to translate, in practical terms, into the ability to randomly select among moves in any given information set at the moment when an information set is actually encountered. Consequently, because a mixed strategy implies a unique behavior strategy in games of perfect recall, it is generally thought that one may use, with equal confidence and correctness, either behavior strategies or mixed strategies in these games. To put this in slightly different language, mixed strategies and behavior strategies are thought to be operationally equivalent in games of perfect recall. The formal expression of that equivalence is due to Kuhn (1953).

2.3. The Nash Equilibrium

The most pervasive concept of equilibrium in games is the *Nash equilibrium*, also known as the *noncooperative equilibrium* and as the *equilibrium point*. For a vector $s \in S$, let $s\backslash t_i = (s_1, \cdots, s_{t-1}, t_i, s_{i+1}, \cdots, s_n)$. An *equilibrium point* is a strategy combination $s^* \in S$ such that $P_i(s^*) \geq P_i(s^*\backslash t_i)$ for all $t_i \in S_i$ and all $i \in N$. At a Nash equilibrium, each player is using a strategy that maximizes her payoff, given the strategies of the other players. The famous existence result of Nash (1951) is

Theorem (Nash 1951). Any finite game has an equilibrium point.

An important generalization of the Nash theorem, due to Nikaido and Isoda (1955), is

Theorem (Nikaido-Isoda 1955) Any game $\Gamma = (N, S, P)$ in which n is finite, $S_i \in \mathbf{R}^m$ is compact and convex, and P_i is continuous in s and quasiconcave in s_i has a noncooperative equilibrium.

The Nikaido-Isoda theorem is important because it applies to games with uncountably infinite strategy spaces; for example, oligopoly games in which firms can choose prices and/or output levels anywhere in some interval of real numbers.

An analytical construct called the *best reply mapping* is often used in conjunction with noncooperative games. It provides a natural alternative way to express equilibrium and is also used directly in framing sequential equilibrium. In a game $\Gamma = (N, S, P)$ the *best reply mapping for player i* is $r_i(s) = \{u_i \in S_i | P_i(s\backslash u_i) \geq P_i(s\backslash t_i)$ for all $t_i \in S_i\}$. That is, $r_i(s)$ is a subset of S_i consisting of the payoff maximizing strategies for player i, given that the other players are choosing $s_j (j \neq i)$. The *best reply mapping* is $r(s) = \times_{i \in N} r_i(s)$. s^* is an equilibrium of Γ if and only if $s^* \in r(s^*)$; that is, if and only if s^* is a fixed point of r.

3. Refinements of the Nash Equilibrium

Refinements of the Nash equilibrium are just what they sound like; Nash equilibria that satisfy some additional condition or conditions. Refinements are motivated by either or both of two considerations. The first is that some Nash equilibria in some games appear flawed as rationally acceptable outcomes. The second is that predictability is increased if the field of equilibria is narrowed and it is at its sharpest when there is only one (surviving) equilibrium. The earliest refinements, subgame perfection, perfection, and proper equilibrium, appear motivated primarily by the first consideration; that of eliminating noncredible equilibria. Sequential equilibrium comes very close to being a restatement of perfect equilibrium in an operationally more useful and appealing form. An impressive cottage industry in refinements has arisen with many of the new contenders seeming to be dreamed up to yield a unique outcome in a special context.

3.1. Subgame Perfect Equilibrium

Subgame perfect equilibrium is the first, and still perhaps the most important, refinement. Many games have *subgames*. In effect, a subgame is the continuation of a game, from a sepcific decision node on to the end of the game, when the continuation satisfies all of the conditions required of games. A strategy combination is subgame perfect if, on each subgame, the continuation of the strategy combination is a Nash equilibrium. A Nash equilibrium can be an intuitively unconvincing way for the players to behave because one of the players uses a strategy that has him act against his interest *at an informations set that will not be reached*. In figure 2–5 the strategy combination left (L) for player 1 and, at each information set, left (l) for player 2 is such a Nash equilibrium. Given player 2's strategy of (l, l), L is certainly the best choice for player 1; however, (l, l) would not be rational for player 2 if player 1 actually chose C or R. One could say that the equilibrium (L, l, l) is supported by a noncredible threat by player 2.

Before explaining subgame perfection, subgames must first be more precisely described. Let $\Gamma = (N, Y, p, I, q)$ be an extensive form game and let $d \in D$ be any decision node. The *subtree* Y_d is the part of the tree Y that includes the node d and all nodes that follow d. In a like manner, D_d is the set of decision nodes that are part of Y_d, T_d is the set of terminal nodes of Y_d, p_d is the payoff function restricted to T_d, I_d is the collection of information sets that can be reached starting from d, and q_d is the collection of probability distributions for nature applying to nature's

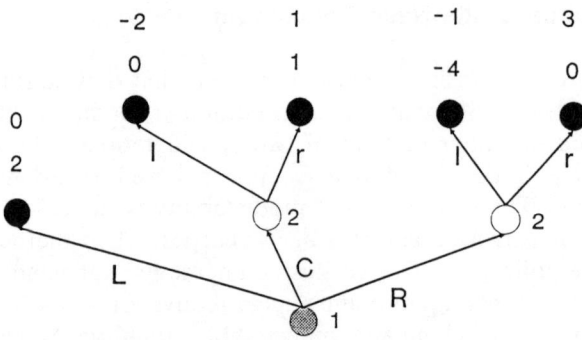

Figure 2-5. Subgame Perfection

information sets within Y_d. Does (N, Y_d, p_d, I_d, q_d) constitute a subgame of Γ? The answer depends upon one fundamental condition. (N, Y_d, p_d, I_d, q_d) is a subgame if *every decision node in each information set contained in I_d lies in the tree Y_d*. Thus in figure 2–3, the remainder of the game from the node d constitutes a subgame, because every information set encountered in Y_d has all of its nodes contained in Y_d. But the remainder of the game from d' is not a subgame, because there is an information set of player 3 that is only partially contained in $Y_{d'}$. Similarly, the remainder of the game from d'' is not a subgame, because d'' itself belongs to an information set that does not lie wholly within $Y_{d''}$. A game Γ is a subgame of itself, so every game has a subgame, but not all games have *proper subgames* (i.e., subgames that differ from Γ).

Suppose that $\Gamma_d = (N, Y_d, p_d, I_d, q_d)$ is a proper subgame of Γ and let π be a pure strategy combination for Γ. Then π_d is the *strategy combination induced by π on Γ_d*. That is, π_d is the portion of π that applies on the part of Γ that is included in the subgame Γ_0. For example in figure 2–3 player 2 has three information set, $I_{21} = \{d\}$, $I_{22} = \{d^*\}$, and $I_{23} = \{d^{**}\}$. At each information set the possible moves are *left* (L) and *right* (R). On the subgame Γ_d that starts at node d, the strategy $\pi_2 = (R_1, L_2, R_3)$ induces the strategy $\pi_d = (R_1, R_3)$.

In parallel to π_d, if s is a mixed strategy combination for Γ, then s_d is the strategy combination induced by s on Γ_d. In figure 2–3, player 2 has eight pure strategies. Suppose s_2 is the mixed strategy that assigns probability as follows: .1 on $(L_1, L_2, L_3,)$, .2 on $(L_1, R_2, L_3,)$, .3 on $(L_1, R_2, R_3,)$, and .4 on $(R_1, R_2, L_3,)$. In the subgame beginning at node d, player 2 has four pure strategies, $(L_1, L_2,)$, (L_1, R_2), (L_1, R_2), and (R_1, R_2). The respective probabilities induced on these four strategies are, then, .3, .3, .4, and 0. More formally, to derive s_d, proceed player by player as follows: $\Pi_{di} = (1, \cdots, m_{di})$ is the strategy set of player i in Γ_d. Each strategy in Π_{di} is the continuation in Γ_d of certain strategies in Π_i. Denote by $\Theta_{dij} \subset \Pi_i$ the set of strategies in Π_i whose continuation in Γ_d corresponds to strategy $j \in \Pi_{di}$. Thus the $\{\Theta_{dij}\}$ ($j = 1, \cdots, m_{di}$) form a partition of Π_i. Keep in mind that, for any player with information sets in Y_d, each strategy $\pi_i \in \Pi_i$ has a continuation in Γ_d. This merely means that each strategy contains directions about how to choose at information sets in Y_d. Consequently, $s_{dij} = \Sigma_{k \in \Theta_{dij}} s_{ik}$.

Given a game $\Gamma = (N, Y, p, I, q)$ and some strategy combination $s \in S$, it is possible to determine whether s, confined to a particular subgame Γ_d, results in equilibrium play on that subgame. If it does, we say s *induces an equilibrium on* Γ_d. In a game Γ, s is a *subgame perfect equilibrium* if it induces an equilibrium on every subgame of Γ. An example is provided in figure 2–5. A pure strategy combination will be written (A, b, c) where the first entry (A) is the strategy of player 1 (L, C, or R), the second entry (b) is the choice of player 2 at her left node (l or r), and the third entry (c) is her choice at her right node. So a strategy for player 2 is (b, c). The following are all equilibria of the game: (L, l, l), (C, r, l), (R, l, r), and (R, r, r). Only (R, r, r) is subgame perfect. At the subgame starting at the right node of player 2, the only equilibrium is (r) and at the subgame starting at the left node, the only equilibrium is (r). Thus $s_2 = (r, r)$ is the only strategy of player 2 that is consistent with subgame perfection. Given (r, r) for player 2, player 1 does best with R. But (L, l, l) is an equilibrium, though it is not subgame perfect; given L, player 2 cannot obtain a higher payoff from any other strategy and given (l, l), player 1 cannot obtain a higher payoff from some other strategy. If there were communication before choosing strategies, player 2 might wish to convince player 1 to choose L by threatening to choose (l, l) if player 1 chooses other than L. Such a threat would be made to induce the choice of a particular equilibrium that player 2 favors; however, the threat is not credible. Player 1 can easily see that if he chooses C or R, then it will not be in the interest of player 2 to carry out her threat; she just loses by doing so. That is, the Nash equilibrium (L, l, l) is supported by a noncredible threat and, for that reason, is not a convincing way to play the game. Requiring equilibrium to be subgame perfect eliminates such equilibria in figure 2–5 and in some other games. Note that any Nash equilibrium induces a Nash equilibrium on subgames that are actually reached (or have a positive probability of being reached) under the equilibrium strategy combination.

Figure 2–6 shows almost the same game as figure 2–5, but subgame

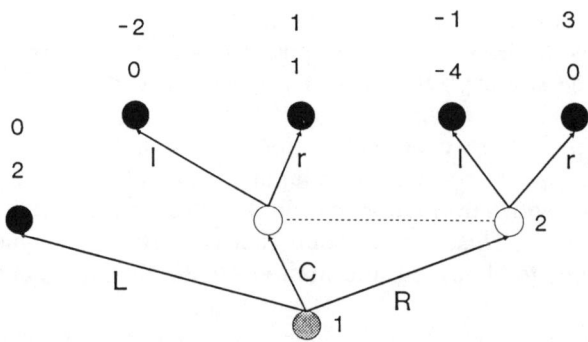

Figure 2–6. Perfect Equilibrium

perfection does not help in the figure 2–6 game. Here player 2 has only one information set and the equilibria of the game are (L, l) and (R, r). Because this game has no proper subgames both equilibria are subgame perfect, but (L, l) is subject to the same criticism as the (L, l) equilibrium in figure 2–5. Namely, that if the information set of player 2 is actually reached, then the only rational choice for her is r. Perfect equilibrium is able to cope with this "unreasonable" equilibrium.

3.2. Perfect Equilibrium

The game in figure 2–6 makes it clear that subgame perfection is not sufficient to rule out all noncredible threats. Subgame perfection only requires a player's choice to be optimal when viewed from information sets at which subgames begin. Perfect equilibrium requires that a player's strategy be optimal when evaluated from any information set at which that player moves. To see how this is accomplished, consider a strategy combination s^* that places strictly positive probability on each pure strategy of each player. Such a strategy, requiring $s_i^* \gg 0$ for player i, is called *completely mixed*. Because s^* is completely mixed, there is a positive probability of reaching any given information set. Therefore, if s^* were also a Nash equilibrium, it would induce payoff maximizing behavior from each information set in the game. This fact is used to define *perfect equilibrium*. The technique is to associate small errors (sometimes called *trembles*) with the players' choices. For player i, suppose that the probability of selecting her jth pure strategy would be $\varepsilon_{ij} > 0$ if she intended to avoid that strategy.

So ε_{ij} is the probability that player i chooses her strategy j by mistake. Thus $\varepsilon_i = (\varepsilon_{i,1}, \cdots, \varepsilon_{i,m_i})$ is a vector of *error probabilities* that are minimum probabilities of selecting the various pure strategies. With these error probabilities in effect, perhaps due to a motor control problem of the player, the effective strategy space for player i is $S_{\varepsilon i} = \{s_i \in S_i | s_i \geq \varepsilon_i\}$. Any strategy from S_i can be played as long as it assigns at least the error probabilities to the various pure strategies. Let $\varepsilon = (\varepsilon_1, \cdots, \varepsilon_n)$ and $S_\varepsilon = \times_{i \in N} S_{\varepsilon i}$. A game with these error probabilities can be given as $\Gamma_\varepsilon = (N, S_\varepsilon, P)$.

An equilibrium s^* of $\Gamma = (N, S, P)$ is a *perfect equilibrium* if there exists an error sequence $\{\varepsilon^k\}$ converging to the zero error sequence such that s^* is the limit of equilibria of the sequence of games $\{\Gamma_{\varepsilon^k}\}$. That is, suppose $s^*_{\varepsilon^k}$ is an equilibrium of $\Gamma^*_{\varepsilon^k}$, that $\{\varepsilon^k\}$ converges to zero errors as $k \to \infty$, and that $s^*_{\varepsilon^k}$ converges to s^*. Then s^* is a perfect equilibrium of Γ. It is not necessary that all sequences $\{s^*_{\varepsilon^k}\}$ converge to s^* or that some sequence converge for every convergent error sequence $\{\varepsilon^k\}$. It is only required that there exist a convergent error sequence and related convergent equilibrium sequence that converges to s^*.

3.3. Sequential Equilibrium

When a player selects a Nash equilibrium strategy, the assumption that the player is rational implies something about that player's beliefs. Suppose, for example, that s^* is an equilibrium. Justifying that player i will choose s^*_i is typically based on claiming that player i believes player $j(j \neq i)$ will choose s^*_j. However, the definition of Nash equilibrium makes no mention of beliefs and has no explicit requirement that the behavior of a player be predicated on the beliefs that other players hold about her. Sequential equilibrium, developed for finite games, is defined to explicitly model each player's beliefs at each information set at which the player moves, to require consistency between the beliefs a player has about other players' behavior and their actual behavior, and to require behavior for each player that is optimal in the light of the player's beliefs. Sequential equilibrium is nearly equivalent to perfect equilibrium and consists of a *system of beliefs* (λ) plus a behavior strategy combination (β); each β^i is a best reply for player i, relative to his beliefs, λ^i, and his beliefs must possess a natural consistency property. The pair (λ, β) is called an *assessment*.

More precisely, a *system of beliefs* for the players in Γ is a family of probability distributions, denoted λ, where $\lambda = (\lambda^1, \lambda^2, \cdots, \lambda^n)$, $\lambda^i = (\lambda^i_1, \lambda^i_2, \cdots, \lambda^i_{r_i})$, and each $\lambda^i_j = (\lambda^i_{j1}, \lambda^i_{j2}, \cdots, \lambda^i_{jm_{ij}})$ is a probability distribution over the nodes in I_{ij}. Consequently $\lambda^i_{j\ell} \geq 0$ and $\sum_{\ell=1}^{m_{ij}} \lambda^i_{j\ell} = 1$. β^i is a *sequential best reply to* (β, λ) *for player* i if β^i maximizes the payoff of player i from each information set $I_{ij} \in I^i$, given the beliefs, λ^i of player i. β is a *sequential best reply* to (β, λ) if it is a sequential best reply for each player $i \in N$.

Consistency is somewhat more involved to explain. Imagine first that β is completely mixed. Then β implies a unique system of beliefs, λ, that is calculated from β by using β to determine the probability of selecting any move in a particular information set, given the attainment of that information set. If β is not completely mixed, then λ is consistent with β if there exists a sequence $\{\beta(k)\}$ of completely mixed behavior strategy combinations that converge to β, with $\{\lambda(k)\}$ being the companion sequence of unique consistent beliefs, and with λ being the limit of $\{\lambda(k)\}$. To summarize, let $\Gamma = (N, Y, p, I, q)$ be a finite extensive form game of perfect recall and let (β, λ) be an assessment. Then (β, λ) is a *sequential equilibrium* if λ is consistent with β and β is a sequential best reply to (β, λ). Kreps and Wilson (1982) prove that a perfect equilibrium s^* in a finite game of perfect recall is a sequential equilibrium in the sense that there exists an assessment (β, λ) such that β is the unique behavior strategy combination induced by s^* and (β, λ) is a sequential equilibrium. The converse does not hold in general; if (β, λ) is a sequential equilibrium, β need not be a perfect equilibrium.

The divergence between perfect and sequential equilibrium is related to their treatment of weakly dominated strategies. A perfect equilibrium cannot include weakly dominated strategies, while a sequential equilibrium can do so. Both require players to use strategies that prescribe optimal (best reply) behavior by each player from each of the player's information sets; thus both rule out any weakly dominated strategies that would permit a player to follow a nonoptimal strategy on a subgame that lies off the equilibrium path. In a measure-theoretic sense, most sequential equilibria are also perfect.

4. Rationality, Selection, and Prediction

If the Nash equilibrium were unique in all games of interest, then much less attention would be paid to refinements of equilibrium; with few exceptions, there would be general agreement that "rationality" required playing the unique equilibrium. However, the Nash equilibrium is not unique in many games, leading to various proposals for refining or abandoning it. The refinements covered above are examples of suggestions aimed at both reducing the number of equilibria that are taken seriously and eliminating some equilibria that are based on beliefs that do not appear to be reasonable beliefs for players to hold. These refinements use *backward induction*. In addition, one could bring in *forward induction, Pareto dominance*, and *preplay communication* (also known lately as *cheap talk*) to eliminate some candidate equilibria. These will be examined in turn. A tack that goes in the reverse direction is *rationalizability* where the point is to determine the largest set of strategies that might possibly be consistent with reasonable beliefs on the part of a player. While other criteria seek to pare down the set of Nash equilibria by bringing in more stringent requirements for "rational" play, *rationalizable strategies* takes the view that any strategy of a player is "rational" to use if it is a best reply to something that might possibly be used by the other players. After the more restrictive concepts are examined, we will look at rationalizability.

4.1. Backward Induction

The principal of backward induction is primarily familiar from finite horizon decision theory. In a decision setting where one person must formulate a policy that determines an action in each of T periods of time, it is customary to solve for the optimal policy in steps, starting with the final period, T. The basic idea is that "acceptable" behavior is determined for the last time period, then acceptable behavior is determined for period $T - 1$ conditional on the previously calculated acceptable behavior being followed at period T. In general, acceptable behavior is found for each period t on the assumption that acceptable behavior will be followed in all later periods. In decision theory, there is only one agent and that agent is assumed to be maximizing a utility function. The one stage problem is often easily solved as a function of the information that will have become available at that time. If such information can be summarized in the form of a *state variable* x_T, then some policy ϕ_T will assign a decision for period T as a function of x_T. Having found the optimal ϕ_T, the optimal policy functiont ϕ_{T-1} can be found as a function of x_{T-1} on the assumption that ϕ_T will govern choice at time T. Working backward, at any time t, the optimal $\phi_t(x_t)$ is found on the assumption that choices in the later periods will be given by $\phi_\tau(x_\tau)$, $\tau = t + 1, \cdots, T$.

Subgame perfection, perfection, and sequential equilibrium are based on backward induction, though in slightly different ways. For subgame perfection, backward induction is put to work by first identifying all subgames that have no proper subgames. Call these *stage 1 subgames*.

Acceptable behavior in the stage 1 subgames is Nash equilibrium behavior. Next stage 2 subgames, that is, those that have only stage 1 subgames as proper subgames, are examined. Acceptable behavior in stage 2 subgames is Nash equilibrium behavior, subject to the condition that acceptable behavior will be followed in any later subgames (i.e., in the stage 1 subgames). The procedure continues inductively in the obvious way. For perfect equilibrium and for sequential equilibrium, backward induction is used to identify acceptable behavior from each information set on the a condition that acceptable behavior will be followed from every subsequent information set.

4.2. Forward Induction

With respect to backward induction, the optimal choice at time t is affected by the choices that are believed to be optimal for later times. Forward induction is the reverse; the optimal choice at time t is affected by the belief that past choices were part of an optimal overall strategy. Thus that certain actions from today onward, though optimal when viewed from today onward, are not consistent with optimality of previous choices and are consequently regarded as unreasonable. This is illustrated in figure 2-7, which shows a variant of the *battle of the sexes*. H is for *husband* and W is for *wife*. If the choice marked B were omitted, the game would be the standard battle of the sexes in which the two players must simultaneously pick an activity for Saturday night among concert (C) and football (F). If they pick the same thing, they carry out that choice, but if they pick different things, they stay home. They both rank staying home below the other two alternatives, but he prefers F to C, while she prefers C to F. The payoffs are symmetric; there are two pure strategy equilibria (F, F) and (C, C). There is also a mixed strategy equilibrium but that will not concern us here.

The game in figure 2-7 differs by presenting H with another choice; he can go bear hunting with the boys (B), which yields him higher utility than the concert (C, C). Forward induction can be used in this game to point out that 1) H prefers the outcome from bear hunting (B) to the concert, 2) if W actually reaches her information set, then she knows that H did not choose bear hunting, 3) if H did not choose bear hunting, then he must expect to obtain a higher payoff than he can get by bear hunting, and 4) therefore, H must have chosen football, as (F, F) is the only (pure) strategy combination that gives H a higher payoff than he obtains by choosing bear

A REVIEW OF REFINEMENTS, EQUILIBRIUM SELECTION

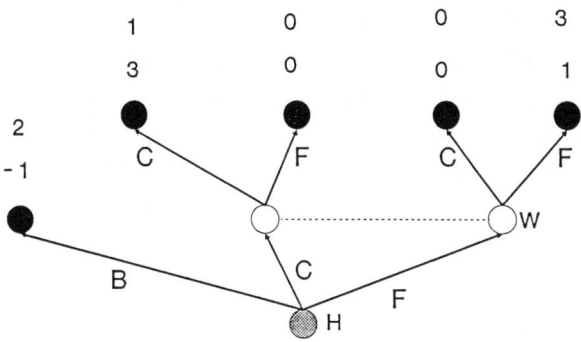

Figure 2–7. Forward Induction

hunting. At *W*'s information set, she reasons that *H* choose *F* as part of an optimal plan that will yield him more than he could have had from the rejected alternative, *B*. Put another way, the rejection of *B* is taken as a signal by *W* concerning the choice made between *C* and *F*. Essentially the same game could be represented slightly differently. Suppose that *H* chooses *B* or not *B*, and that this choice is observed by *W*. Then *W* picks *C* or *F*, but *H* does not see the choice she has made. Finally, *H* chooses between *C* and *F*. In this version, if *W* reaches her information set, she knows that *H* has turned down the bear hunt, but *H* is not yet committed to either *C* or *F*. The reasoning outlined above still holds. Forward induction suggests that *H* is fully intending to choose *F because he is following a plan that is optimal from the start*; therefore, he would not turn down the bear hunt unless he expected to do better. Nonetheless, a subgame begins with *W*'s information set and both (*F*, *F*) and (*C*, *C*) are equilibria of this subgame.

Strategies like the bear hunting strategy of *H* are called *outside options* because they are alternatives that remove a player from interaction with the other players in the game. The interpretation that forward induction brings to the game when *H* declines the outside option is appealing; however, it is easy to construct games in which forward induction arguments like the one above become ambiguous. Suppose the game is changed so that both *H* and *W* have outside options giving them more than one unit of payoff. What should we conclude when *both* players forgo their outside options? That they both expect to have their way in the ensuing battle of the sexes? Clearly these beliefs are inconsistent and forward induction will not select or in any way reduce the scope of rational behavior in the remainder of the game.

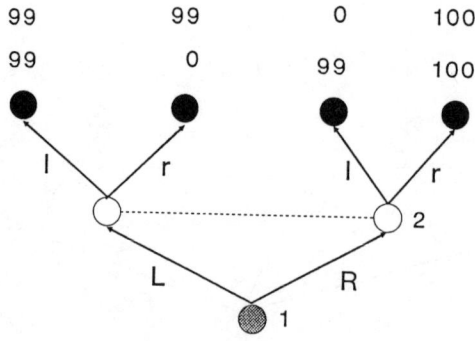

Figure 2-8. Pareto Dominance and Risk

4.3. Pareto Dominance

The *Pareto dominance* principle is very simple; it says that any Nash equilibrium that is Pareto dominated by another Nash equilibrium should be ignored. If that were a widely accepted principle, it would yield a unique outcome in games in which one Nash equilibrium was Pareto superior to all others. Figure 2-8 provides an example with two Nash equilibria (R, r) and (L, l), with (R, r) being Pareto dominant. If both players lack total confidence that the other player will choose *right*, then the (R, r) equilibrium may not be seen. From the standpoint of player 1, if she thinks the probability of player 2 choosing l is greater than 0.01, then she will want to choose L. A guarantee of 99, no matter what player 2 chooses, looks very good when choosing R and failing to achieve equilibrium will bring a payoff of zero. The safety of (L, l) as compared with (R, r) comes under the heading of *risk dominance*, a concept defined, discussed, and used by Harsanyi and Selten (1988).

4.4. Cheap Talk

Preplay communication, the ability of the players to hold conversations or to send one another messages prior to starting the formal moves of the game, has been referred to in books and articles on game theory for many years. Traditionally, preplay communication is mentioned to say that a) it is not permitted, b) it is permitted, but does not matter, or c) is permitted and will, in some mysterious way, determine which of many equilibria will be selected by the players. In these earlier instances preplay communication has never been part of the formal structure of the game. More recently formal analysis has been undertaken to understand the exact role that preplay communication could have on the outcomes of games. Doing this requires specification of the number of messages that one player can send another, the timing of such messages, any desired restrictions on content, and so forth. Consequently, the messages that can be sent and times when they can be sent are built into the extensive form of the game.

To see the effect that preplay communication could have, consider the battle of the sexes in its usual form (i.e., H has only two choices, C and F). Suppose further that W is able to send one message, but H cannot send any messages. It is likely she would send "I intend to select C" and it is also likely that H would believe this message. The consequence would be that the (C, C) equilibrium would be chosen. An issue in the cheap talk literature concerns the believability of messages; after all, messages are not binding and they cost nothing to send. In the battle of the sexes example, the message is believable because it appears in the interest of W to carry out the action she announces in her message. That is, if her message were believed by H, then her best reply would be to act as the message claims she will act. Contrast this with figure 2–5. If there were preplay communication and player 2 said to player 1, "If you choose C or R, then I will choose ℓ," player 1 would likely scoff and disbelieve, because it would not be in the interest of player 2 to choose ℓ after player 1 has chosen C or R.

4.5. Rationalizability

To this point, Nash equilibrium has been discussed and various ways of reducing the set of Nash equilibria have been considered. When W says she will choose C, we might say it is *rational* for H to believe her and act accordingly. When we see that the equilibrium (L, l, l) in figure 2–5 requires that player 2 choose suboptimally on unreached subgames in order for L to be a best reply for player 1, we may say the equilibrium rests on an *irrationality*. Such considerations all appear to take Nash equilibrium behavior for granted as part of rational behavior. That is, they appear to start from an assumption that Nash equilibrium behavior is a necessary implication of rational play. Suppose that assumption is explicitly abandoned and the question is posed: What strategies can be ruled out and what strategies are left in consideration when we regard as rational any strategy that is a best reply for a player to those strategies of the other players that are also best replies?

To be more precise, let $\Gamma = (N, S, P)$ be a game in strategic form and suppose $T_i \subset S_i$ for each $i \in N$. For $t = (t_1, \cdots, t_n) \in \times_{i \in N} T_i \equiv T$, each strategy t_i is *rationalizable* if $t_i \in r_i(u^i)$ for some $u^i \in T$. That is, player i regards T as the strategy space from which she believes all players will select their strategies. Each strategy of her own in T_i is a justifiable choice because it is a best reply to something in T (i.e., to something she believes the other players might select), and each $t_j \in T_j$ is a strategy that player i regards as a justifiable choice for player j because t_j is a best reply to something in T. Thus no $t \in T$ can be ruled out as being irrational because each $t_i \in T_i$ ($i \in N$) is a best reply to something that appears a possible choice.

Consider the game of matching pennies, with utility measured by money. It is well known that the game has a unique Nash equilibrium at which both players randomize, choosing heads with probability $\frac{1}{2}$. If player 1 believes player 2 will choose heads with probability $\frac{1}{2}$, any mixed strategy of player 1 is a best reply, including choosing heads with certainly or tails with certainly. A parallel statement can be made for player 2. In this game, the whole of S (the set of mixed strategy combinations) is rationalizable; for any $s \in S$, s_1 is a best reply to something in S and similarly for s_2.

The complete set of rationalizable strategies can be specified in the following way. For $A \subset S$, let $\lambda_i(A) = \cup_{s \in A} r_i(s)$ and $\lambda(A) = \times_{i \in N} \lambda_i(A)$. Thus $\lambda(A) \subset S$ is the set of strategy combinations that are best replies to something in A. Denote by $\lambda^k(A)$ the k-fold composition of λ with itself. Thus $\lambda^1 \equiv \lambda$ and $\lambda^k(A) \equiv \lambda[\lambda^{k-1}(A)]$. The set of all rationalizable strategies is $\lim_{k \to \infty} \lambda^k(S) \equiv R(S)$. The set $R(S)$ is arrived at by the successive elimination of strategies that are not best replies to something that has not yet been eliminated. $R(S)$ has the structure of a Cartesian product. That is, there is $R_i(s)$ for player i such that $R(S) = \times_{i \in N} R_i(S)$. The Bernheim and Pearce argument is that rational behavior only requires that player i choose a strategy from $R_i(S)$.

When might we expect players to choose any of the rationalizable strategies? A real answer to this question awaits extensive empirical (experimental) research; however, to hazard a guess, rationalizability is likely to be important when other selection criteria all fail and when preplay communication is not available. If, at one extreme, a game has a unique Nash equilibrium and, at that equilibrium, each player's best reply is unique, such an equilibrium strikes me as very likely to be chosen. Some games will have a focal equilibrium, some a Pareto dominant equilibrium that is not especially risky, some an equilibrium that uniquely emerges from the only "reasonable" forward induction argument that might be made, etc. Yet another possibility is a game that falls into a standard pattern and is played by various people at various times, with each instance of the game being unconnected with all others. An example occurs when two cars are driving toward one another in an empty parking lot. That essentially the same game is played independently on many occasions by people permits a *convention* to arise (pass on the right) that selects a particular Nash equilibrium. If all these possibilities fail, then perhaps any rationalizable behavior is reasonable to expect and has a chance of being seen.

5. Repeated Games

Repeated games are of interest as a consequence of two common observations. First, the play of a many games involves continued interaction among a fixed set of players who are known to one another. Second, many situations of interest to economists and others have a stationary, repetitive structure that looks as if a particular game is being played over and over again. Perhaps the most conspicuous example is oligopoly. On one hand, oligopoly has been studied mainly as a static, one shot game; firms choose output levels (or perhaps prices) once-for-all, and this determines prices and profits. On the other hand, it has been long recognized that this *single-period oligopoly game* is played over and over again. The format reviewed below is that a single shot game (N, S, P) is played by the unchanging set N of players at time $t = 0$, $t = 1$, and so forth forever. The sets S_i will be interpreted as sets of pure strategies. If $s_t \in S$ is chosen by the players at time t, then the *repeated game payoff* to player i is $\Sigma_{t=0}^{\infty} \alpha^t P_i(s_t)$. Payoff outcomes that could never be Nash equilibrium outcomes in the single shot can be subgame perfect Nash equilibrium outcomes in the repeated game. For example, Pareto optimal payoffs that do not coincide with single shot Nash payoffs can be achieved in the repeated game. The subsections that follow will first, set up the repeated game model, second, examine trigger strategy equilibria, third, look at the Fudenberg-Maskin extension of the "folk theorem for repeated games," and, fourth briefly consider games in which the past actions of other players are imperfectly observable (*imperfect monitoring*).

5.1. The Repeated Game Model

In an infinitely repeated game based upon (N, S, P), it is vital to keep in mind that *the game* is not (N, S, P), but is actually the repeated game; a game in which the strategy of a player is a sequence of decision rules, one for each time period. The game (N, S, P) is referred to as the *stage game* or the *constituent game*. At each time $t > 0$, player i will choose s_{it} as a function of the observed previous actions of all players, that is, as a function of $(s_0, \cdots, s_{t-1}) \in S^t$. In general, player i can use different decision rules in different time periods and *perfect monitoring* is assumed. Perfect monitoring means that, at any time $t > 0$, all players know the actual $s_\tau (\tau < t)$. If f_{it} were the decision rule used at time t, then a repeated game strategy for player i has the form $\sigma_i = (s_{i0}, f_{i1}, f_{i2}, \cdots)$. Denote by $\mathscr{S}_i$ the set of such strategies, define $\mathscr{S} = \times_{i \in N} \mathscr{S}_i$, and let $f_t = (f_{1t}, \cdots, f_{nt})$. A given pure repeated game strategy combination $\sigma \in \mathscr{S}$ will induce a specific path of actions (single shot game strategy combinations) over time, denoted $u_t(\sigma)$ for $t = 0, 1, \cdots$, where $u_0(\sigma) = s_0$, $u_1(\sigma) = f_1(u_0(\sigma)) \in S$, and, recursively, $u_t(\sigma) = f_t(u_0(\sigma), \cdots, u_{t-1}(\sigma)) \in S$. The repeated game payoff function of player i, expressing payoffs as a function of repeated game strategies, is $G_i(\sigma) = \Sigma_{t=0}^\infty \alpha^t P_i[u_t(\sigma)]$. Letting $G = (G_1, \cdots, G_n)$, the repeated game is given by $(N, \mathscr{S}, G)$.

5.2. Trigger Strategy Equilibria

Suppose that the stage game (N, S, P) has a unique equilibrium, $s^c \in S$ and that $P(s^c)$ is not Pareto optimal (there exists $s' \in S$ such that $P_i(s') > P_i(s^c)$ for all i). Then, depending on the value of α, there may be relatively simple strategy combinations for the repeated game that are subgame perfect equilibria and at which the payoffs $P(s')$ are achieved in each period. The strategies to be considered are called *trigger strategies* and their structure is this: For player i, σ_i calls for a) $s_{i0} = s'_i$, b) for $t > 0$, $s_{it} = s'_i$ if $s_\tau = s'$ for $\tau = 0, \cdots, t-1$, but $s_{it} = s^c_i$ otherwise. Stated in the language of expectations concerning tacitly cooperative behavior, player i will begin by choosing s'_i and she expects the other players to cooperate by choosing s'_j ($j \neq i$); player i will continue to choose s_i' unless some player $j \in N$ defects by choosing differently from s'_j; and upon seeing a defection, player i will switch to choosing according to the single shot equilibrium, that is s^c_i. Players will cooperate in choosing s' but will *trigger* to s^c if any break in cooperation occurs. Such strategies are called *grim trigger strategies* in contrast with strategies that trigger to s^c for some finite number of periods and then return to s'.

To see when such a strategy combination is a subgame perfect equilibrium, consider first what constitutes a subgame in the repeated game. Thinking of the game in extensive form, in each period t player 1 chooses first, player 2 chooses next but does not know what player 1 selected, player 3 chooses next but does now know what player 1 or 2 selected, and so forth. After player n chooses, all period t choices become known to all players. As the game is one of perfect recall, it is clear that a subgame begins at each information set where player 1 moves, but no subgames begin at any other information sets. Clearly, the information sets of player 1 contain exactly one node, while those of all other players contain more than one node. In addition, any information set that can be reached following an information set of player 1 contains only nodes that also can be reached from that information set of player 1. These conditions ensure that subgames begin at, and only at, information sets of player 1. Put in simpler language, a new subgame begins at any moment when a new time period starts; however, at any time t, there are as many subgames as there are histories of play, $h_t = (s_0, \cdots, s_{t-1}) \in S^t$.

Now suppose that $\sigma \in \mathscr{S}$ is a trigger strategy combination based upon (s^*, s^c) and satisfying $P(s^*) \gg P(s^c)$. In considering the best reply of player i to σ, player i can stay with σ_i to receive $P_i(s^*)$ in all periods for a discounted payoff of $P_i(s^*)/(1 - \alpha)$ or can follow some alternative strategy. If player i deviates at t from choosing s_i^*, the other players trigger to s_j^c at $t + 1$, so that, from $t + 1$ onward, the best player i can do is choose s_i^c and obtain a payoff of $P_i(s^c)$ per period from $t + 1$ onward. At period t, player i can achieve a payoff of $\phi_i(s^*) = \max_{s_i \in S_i} P_i(s^*\backslash s_i)$. Thus σ is a subgame perfect equilibrium if $P_i(s^*)/(1 - \alpha) \geq \phi_i(s^*) + \alpha P_i(s^c)/(1 - \alpha)$ or

$$\alpha \geq \frac{\phi_i(s^*) - p_i(s^*)}{\phi_i(s^*) - p_i(s^c)} \quad \text{for all } i \in N \qquad (2)$$

Checking for subgame perfection requires that each time t and each possible history $h_t = (s_0, \cdots, s_{t-1})$ be examined; however, the histories sort into two categories. One is the unique history $(s^*, \cdots, s^*)$, which calls for continuing to choose s^* and the other is every other history, which calls for choosing s^c. It is clear that σ is subgame perfect if equation 2 holds because no player i in any subgame can obtain a higher payoff in that subgame by deviating from σ_i.

There are two additional observations to make about trigger strategy equilibria. The first concerns the payoff outcomes, $P(s^*)$, that can be supported by trigger strategies. Clearly, as long as $P(s^*) \gg P(s^c)$, the outcome $P(s^*)$ can be supported if α is sufficiently close to one. If there are multiple equilibria in (N, S, P), then any such equilibrium can be used to play the role of s^c. Thus any attainable payoff that strictly Pareto dominates a single shot Nash equilibrium can be supported if the discount parameter is large enough. The second observation is about the necessity of using *grim* trigger strategies. Is it possible to replace the grim trigger, under which one defection moves the players forever to single shot equilibrium play, with a *finite reversion trigger strategy* under which the players follow a defection with T periods at the single shot equilibrium and then return to s^*? Yes, it is possible if equation 2 holds with strict inequality for all players. Then there would be a finite T that would provide the needed incentives.

5.3. A Subgame Perfect Version of the Folk Theorem

The "folk theorem for repeated games" refers to a result that has been known for several decades, but for which the discoverer is not known. The result is that, in an infinitely repeated game without discounting (i.e., where $\alpha = 1$) any *individually rational* payoff vector of (N, S, P) can be sustained as the Nash equilibrium payoff of the repeated game. A payoff x_i is individually rational for player i if x_i is not less than the payoff the player would receive when all other players were attempting to minimize her payoff. In general, there are individually rational payoffs that are smaller than Nash equilibrium payoffs. Thus, an important aspect of the *folk theorem for repeated games* is that a much larger set of single shot payoff outcomes can be supported, as compared with those that can be supported by means of trigger strategies. Although the *folk theorem* itself is about Nash equilibrium in games without discounting, more recent results extend the theorem to subgame perfect equilibrium in games with and without discounting. The brief treatment below focuses on subgame perfect equilibrium in games with discounting.

Before the theorem is discussed, some background must be developed. A payoff is *individually rational* if it gives to each player a payoff at least as great as that player can guarantee to herself. What a player can guarantee is a *minimax* payoff. That is, think of a two-player zero-sum game in which player i is one player and the remaining players, taken together (denoted $N\backslash\{i\}$), is the second player. The payoff function of player i in this game is $P_i(s)$ and the payoff function for $N\backslash\{i\}$ is $-P_i(s)$. The strategy space of player i is S_i and of $N\backslash\{i\}$ is $\times_{j \neq i} S_j$. The *minimax payoff of player i* is $\min_{s_{N\backslash\{i\}} \in S_{N\backslash\{i\}}} \max_{s_i \in S_i} P_i(s_i, s_{N\backslash\{i\}}) = P(s^i) \equiv v_i$. Clearly no Nash equilibrium payoff can be less than v_i; however, in general Nash equilibrium payoffs will be strictly larger. Letting $v = (v_1, \cdots, v_n)$, and letting $s' \in S$ satisfy $P(s') \gg v$, the *folk theorem for repeated games* states that, in a game with no discounting, there exists a Nash equilibrium strategy combination $\sigma' \in \mathcal{S}$ such that $u_t(\sigma') = s'$ for all t. With no discounting, the payoff functions defined above are inappropriate, because they are unbounded. This can be remedied in various ways; however, the key to proving the *folk theorem* is that any player who deviates from s' can be minimaxed long enough to take away any gains resulting from the deviation, after which the strategies revert to choosing s'. Strategies that do this are not subgame perfect, because the nondeviating players do not have proper incentives to carry out the prescribed punishment.

The particular cure to restore subgame perfection that Fudenberg and Maskin (1986) propose for games with discounting is sketched below, using the language of *paths* due to Abreu (1988). A *path* $\mu^k = (\mu_0^k, \mu_1^k, \mu_2^k, \cdots, \mu_t^k, \cdots)$ where each $\mu_t^k \in S$. That is, a path is an infinite sequence of action combinations. For example, playing the strategy combination σ would generate the path $(u_0(\sigma), u_1(\sigma), u_2(\sigma), \cdots)$. Intuitively, a path is like a strategy combination in which there is no history dependent behavior; that is, the decisions made at time t depend only on the date (t), but do not depend upon any actual previous choices made by players. What Abreu does is define a class of strategy combinations, called *simple strategy combinations*, that are made up of $n + 1$ paths, together with rules for changing from one path to another. One path is followed from the start of the game and each of the other paths is associated with a particular player as that player's *punishment path*. The general rule is that the players at time $T + 1$ remain on the path they were on at time T, unless some player deviated at T. In that case, the players switch to the deviator's punishment path and commence that path at its beginning.

The $n + 1$ paths are denoted $\mu^0, \cdots, \mu^n$. The path μ^0 is the *initial* path and, if no player deviates from choosing according to what the initial path prescribes, the players continue along it forever, choosing μ_t^0 at each time t. The response prescribed by the strategy combination to deviation by some player j at time T is to immediately switch to the path μ^j, the *punishment path of player j*, to start on that path from the beginning of it, and to stay on that path forever, unless a further deviation by some player occurs. Thus, until another deviation occurs, the players will choose μ_t^j at time $T + t + 1$. If a further deviation occurs, then the players switch to the punishment path of the deviator and start from the first entry (e.g., μ_0^k) of the new deviator's punishment path, even if that same player has previously deviated.

The equilibrium paths have a very special, simple structure: $\mu^0 = (s', s', s', \cdots)$ and $\mu^i = (s^i, \cdots, s^i, \omega^i, \omega^i, \cdots)$ where $P_j(\omega^i) = P_j(s')$ for $j \neq i$ and $P_i(\omega^i) = P_i(s') - \varepsilon$ for a very small $\varepsilon > 0$. In μ^i the minimax action combination, s^i, occurs in the first T periods. That is, along the initial path, the same action s' is always chosen and $P(s') \gg v$. On a punishment path for player j, player j is minimaxed for T periods and then the players revert to ω^j which yields prepunishment payoffs to all players except the deviator and yield the deviator ε less than the prepunishment payoff. The requirement that any deviation at any time by any player is met by changing to that player's punishment path, starting it from the beginning, means that any deviator i is met with T periods of being held to v_i, following which play reverts to being nearly where it began at the start of the game.

A strategy combination that uses paths $(\mu^0, \cdots, \mu^n)$ is subgame perfect if the discount parameters of the players are sufficiently near one. To be subgame perfect requires, in practice, that none of the following deviations is profitable for a player i: 1) deviation from the initial path, 2) deviation from the player's own punishment path at any point on that path, and 3) deviation from the punishment path of some other player at any point on that path. For deviation from the initial path to be unprofitable, it suffices that T periods of being minimaxed more than makes up for the one period extra payoff obtained by the deviation. No deviation from one's own punishment path is worthwhile when deviation from the initial path does not pay. For deviation from the punishment of another player to yield a lower payoff than sticking with the punishment is due to the extra ε that the punisher will receive when the minimaxing phase ends. This extra ε in perpetuity must make up for the possibility that player i, when punishing player j, could receive even less than v_i per period in the minimaxing phase. But, if α is near enough to unity, the discounted value of that extra ε can be enough to outweigh any short term gains from deviating from the punisher's role.

5.4. Games with Imperfect Monitoring

The pioneering work on this topic is due to Green and Porter (1984) with important further developments due to Abreu, Pearce, and Stacchetti (1986, 1990). All of the results sketched in sections 5.2 and 5.3 required players to have sufficient information to detect defections with absolute certainly. Indeed, in section 5.3 it was necessary, as well, to be sure which player had defected. It is easy to imagine situations in which players receive useful, but not completely reliable, information about past behavior. An example will illuminate the issues. Suppose an infinite horizon Cournot duopoly in which inverse demand is $p_t = \max\{0, 90, -q_{1t} - q_{2t} + \varepsilon_t\}$ where q_{it} is the output of firm i at time t and ε_t is a random variable distributed uniformly over $[-10, 10]$. When the q_{it} are chosen, ε_t is not known to the firms. With costs of zero, the single period payoff of firm i is $p_t q_{it}$ and the single shot Nash equilibrium is $q_1 = q_2 = 30$. Pareto frontier (expected) profits are obtained when $q_1 + q_2 = 45$. Now suppose that, at any time t, each firm knows its own past output choices and past prices. The firm does not observe the other firm's output and does not observe the value of the random variable.

Consider the possibility of maintaining a trigger strategy equilibrium. If the firms intended to coordinate on choosing $q_1 = q_2 = 22.5$ and would trigger permanently to single shot Cournot behavior if the market price were observed to differ from 45 ($= 90 - 22.5 - 22.5$), then they would surely trigger immediately. To have any hope of sustaining collusion, they must make at least two modifications to their trigger strategies: First, trigger when price is below 45, but not when it is above. Clearly, if the players trigger if price differs from 45, they will almost surely trigger in the first period. As the relevant deviations by players are to overproduce, causing price to drop, the triggering should be based one one-sided deviations. Second, play single-shot Cournot for a few periods and then revert to choosing $q_1 = q_2 = 22.5$. Even if no player ever deviates, the random variable will sometimes be negative, causing a price below 45 to be seen. Common sense suggests that it is wise to provide for a return to collusive output levels, because the firms will sometimes trigger in response to a very low value of ε. As they cannot distinguish between a low ε and cheating, they must use a rule based on price alone; however, they can choose the length of punishment time so that it is the minimum required to make cheating unprofitable for either firm.

An additional question is whether $p_t < 45$ is the optimal criterion for triggering to single-shot Cournot output. Two considerations must be balanced against one another. Consider prices in the interval $(30, 45)$ as candidates for the trigger price, p^T. The lower bound is the single-shot Cournot price. The larger is p^T, the higher the probability the firms will trigger due to a low realization of ε; however, the smaller is p^T, the greater the opportunity for a firm to produce a little to much without being detected. Thus there is probably an optimal value for $p^T \in (30, 45)$ at which these two consideration are balanced on the margin.

Consider the demands made by observability in sections 5.2 and 5.3. In section 5.2 it is sufficient to know whether someone cheated, but it is not necessary to know who cheated. Thus if the random variable is removed from the example discussed above, the firms have essentially perfect opportunities to tell if someone has cheated. If $n = 2$, it is clear who cheated; however, if $n > 2$, it would not be possible to tell who. Trigger strategy equilibria would be viable. But in section 5.3 the equilibrium strategies call for punishments based on the identity of the defector; therefore, merely observing market price is not sufficient information to carry out punishments of the type envisaged by Fudenberg and Maskin. Finally, consider a repeated game in which each person's actions in each period can be observed, but in which the single-shot game (N, S, P) is finite. If the collusive single-shot action combination, s^*, is a mixed strategy combination of (N, S, P) then it will not be possible within a single time period to tell whether a player is cheating. It would be possible to detect systematic cheating over a number of periods, however.

References

Abreu, Dilip. 1988. "On the Theory of Infinitely Repeated Games with Discounting," *Econometrica* 56: 383–396.

Abreu, Dilip, David Pearce, and Ennio Stacchetti. 1986. "Optimal Cartel Equilibria with Imperfect Monitoring." *Journal of Economic Theory* 39: 251–269.

Abreu, Dilip, David Pearce, and Ennio Stacchetti. 1990. "Toward a Theory of Discounted Repeated Games with Imperfect Monitoring." *Econometrica* 58: 1259–1281.

Aumann, Robert J., and Lloyd Shapley. 1976. "Long Term Competition—A Game Theoretic Analysis." Mimeograph.

Bernheim, B. Douglas. 1984. "Rationalizable Strategic Behavior." *Econometrica* 52: 1007–1028.

Binmore, Ken. 1992. *Fun and Games: A Text on Game Theory*. Lexington: Heath.

Farrell, Joseph. 1987. "Cheap Talk, Coordination, and Entry." *RAND Journal of Economics* 18: 34–39.

Farrell, Joseph. 1988. "Communication, Coordination, and Nash Equilibrium." *Economic Letters* 27: 209–214.

Friedman, James W. 1971. "A Non-cooperative Equilibrium for Supergames." *Review of Economic Studies* 38: 1–12.

Friedman, James W. 1990. *Game Theory with Applications to Economics*, 2nd ed. New York: Oxford.

Fudenberg, Drew, and Jean Tirole. 1991. *Game Theory*. Cambridge: MIT Press.

Fudenberg, Drew, and Eric Maskin. 1986. "The Folk Theorem in Repeated Games with Discounting and with Incomplete Information." *Econometrica* 54: 533–554.

Green, Edward, and Robert Porter. 1984. "Noncooperative Collusion under Imperfect Price Information." *Econometrica* 52: 87–100.

Harsanyi, John, and Reinhard Selten. 1988. *A General Theory of Equilibrium Selection in Games*. Cambridge: MIT Press.

Kuhn, Harold. 1953. "Extensive Games and the Problem of Information." In H.W. Kuhn, and A.W. Tucker, (eds.), *Contributions to the Theory of Games*, Vol. 2. Princeton: Princeton University Press.

Kohlberg, Elon, and Jean-François Mertens. 1986. "On the Strategic Stability of Equilibria." *Econometrica* 54: 1003–1039.
Kreps, David, and Robert Wilson. 1982. "Sequential Equilibria." *Econometrica* 50: 863–894.
Luce, R. Duncan, and Howard Raiffa. 1957. *Games and Decisions.* New York: Wiley.
Myerson, Roger. 1991. *Game Theory: Analysis of Conflict.* Cambridge: Harvard.
Nash, John F., Jr. 1951. "Non-Cooperative Games." *Annals of Mathematics* 54: 286–295.
Nikaido, Hukukane, and Kazuo Isoda. 1955. "Note on Noncooperative Convex Games." *Pacific Journal of Mathematics* 5: 807–815.
Pearce, David. 1984. "Rationalizable Strategic Behavior and the Problem of Perfection." *Econometrica* 52: 1029–1051.
Rubinstein, Ariel. 1979. "Equilibrium in Supergames with the Overtaking Criterion." *Journal of Economic Theory* 21: 1–9.
Selten, Reinhard. 1965. "Spieltheoretische Behandlung eines Oligopolmodelles mit Nachfrageträgheit." *Zeitschrift für die Gesamte Staatswissenschaft* 12: 301–324.
Selten, Reinhard. 1975. "Reexamination of the Perfectness Concept for Equilibrium Points in Extensive Games." *International Journal of Game Theory* 4: 25–55.
van Damme, Eric. 1987. *Stability and Perfection of Nash Equilibria.* New York: Springer.

3 COORDINATION IN GAMES: A SURVEY

Gary Biglaiser*

1. Introduction

What does it mean for players to coordinate in a game theoretic context? Coordination can be thought of in many different ways. One is that coordination means players find strategies that are part of the same equilibrium. In a game with multiple equilibria it is easy to imagine players selecting strategies resulting in a disequilibrium outcome, because they do not have the same equilibrium in mind. This can arise in different classes of games. One class is where players have identical preferences over the ranking of equilibria. The problem for the players is to achieve the right equilibrium, when all players agree on what is the best equilibrium. The other class of games has the property that the players disagree on the relative ranking of the equilibria. Coordination may still be important even in this type of games, since while players may not have identical preferences regarding

---

* I would like to thank Joel Sobel for advice on the literature. I am particularly indebted to Jim Friedman for many useful discussions and careful readings of this chapter.

	Player 2	
	L	R
T	1, 1	0, 0
D	0, 0	1, 1

Game 1

Figure 3–1. Game 1

the ranking of equilibria, they may still agree that one equilibrium is better than another. Thus, for example, they may always want to coordinate to attain Pareto efficient equilibria. I will now illustrate these different ideas of coordination through some examples.

In his classic book *The Strategy of Conflict*, Schelling gives many examples of what can be termed pure coordination games with no unique Pareto optimal equilibrium. Here are two of these examples. First, suppose that you and a friend are to meet in New York City on a given day, but neither of you know when or where you are supposed to meet and there is no way for you to communicate with each other. Furthermore, neither of you care about the location or timing of your meeting as long as you find each other. The essence of this example can be captured in the payoff matrix of Game 1 (figure 3–1). In Game 1, both players know that they are supposed to meet in New York City at noon but they do not know which location, where there is only a choice of two, either Times Square or the Empire State Building. Each player gets a utility payoff if they meet of 1, while if they do not meet they receive no payoff.

The second example is one where you and your friend are separated and told that each of you should write down either heads or tails on a piece of paper. If each of you writes down the same side of the coin, then each of you wins a prize; otherwise, each of you receives nothing. This example can also be illustrated using Game 1, where the prize increases each of the players utility by a unit. In both of these examples, players do not care what equilibrium is played, location and time in the first and heads or tails in the second, as long as they succeed in coordinating, go to the same place at the same time or pick the same side of the coin. Thus, there are multiple Nash equilibria in each of the games, with no equilibrium being dominated by any other equilibrium as is illustrated in Game 1.

Another way to view coordination can be seen by examining the game

COORDINATION IN GAMES: A SURVEY

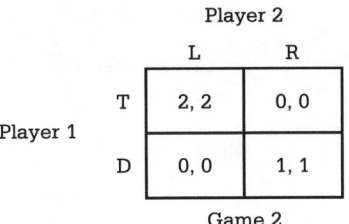

Figure 3–2. Game 2

illustrated in figure 3–2. In Game 2, there are two Nash equilibria, player 1 choosing T and player 2 choosing L and player 1 choosing D and player 2 choosing R. The equilibria can be Pareto ranked, with both players preferring the (T, L) equilibrium to the (D, R) equilibrium. In fact, the (T, L) equilibrium gives both players the highest payoff that they can receive in the game.

In both of the above two ideas of coordination, illustrated by Games 1 and 2, each player wants to maximize the other players's payoff because this will maximize his own payoff. Thus, players have no incentive to play strategically in the sense of trying to fool the other player. But, if they have no way of communicating with each other, they still face a "real" coordination problem. Schelling discussed the use of focal points or clues as a way for players to coordinate their actions or strategies. An example of a focal point in Game 1 is to meet at Times Square at noon. This is because since players are supposed to meet at a given *time* they should coordinate on a location that has time as a feature. In Game 2, equilibrium (T, L) appears focal, since it gives each of the players the highest payoff that they can receive. The notion that players should coordinate using focal points is intuitively appealing. Unfortunately, there still remain unresolved problems with this notion. First, nothing may be uniquely focal. For instance, in the naming of a side of a coin example, neither heads or tails appears focal in the description of the game. The second problem, at least from an economic theorist's point of view, is that formalizing the idea of focal points is quite difficult, see Crawford (1982) for an attempt at this formalization.

Another way to view coordination is by looking at a game such as Game 3 (figure 3–3), which is the prisoners' dilemma. In the single shot version of Game 3, it is a dominant strategy for player 1 to choose D and player 2 to play R. Thus, coordination in this game seems meaningless,

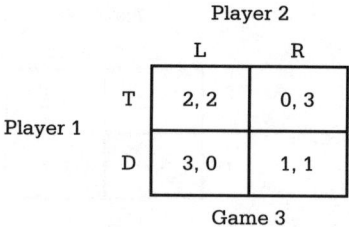

Figure 3–3. Game 3

since each player's optimal strategy is independent of the other player's strategy choice. On the other hand, if players play this game an infinite number of times, other per period equilibrium payoffs can also be supported. In fact, from the folk theorem, see Friedman (1971) and Fudenberg and Maskin (1986) as examples, we know that for sufficiently low discount rates any feasible individually rational payoff can be supported as a subgame perfect equilibrium. A payoff is individually rational if it is not smaller than a payoff that a player can guarantee to herself. In Game 2, each player can always guarantee herself a per period payoff of at least 1. In games with some divergence of interests between the players, but with possibilities for cooperation or coordination, such as the infinitely repeated prisoners' dilemma, new issues arise concerning coordination. Take figure 3–4, which illustrates the set of feasible per period payoffs in the infinitely repeated play of Game 2 as the discount rates of both players go to 0. By the folk theorem, as the discount rates go to 0, any pair of payoffs that are in the outlined region can be supported as an equilibrium of the game. Coordination involves two ideas in this game. First, players must coordinate their strategies on a given equilibrium, just as in the earlier examples of coordination. Second, take the pair of payoffs (1.5, 1.5) denoted by letter A in figure 3–4. This payoff pair could be an equilibrium of this game. Of course, this pair is not Pareto efficient. Both players would prefer to coordinate on different strategies where each can receive higher payoffs. For example, any payoff pair on the utility possibility frontier, which is the set of Pareto efficient outcomes, giving each player at least 1.5 would be strictly preferred by both the players to the putative equilibrium.

In the rest of this chapter, I will review a few articles that deal with these ideas of coordination. In section 2, the theory papers are presented. I first discuss Crawford and Haller (1990). In the main example of their paper, players face a repeated pure coordination game with many unranked Pareto optimal equilibria. This example is in the same spirit as the ones that were earlier attributed to Schelling; in particular it is Game 1. The

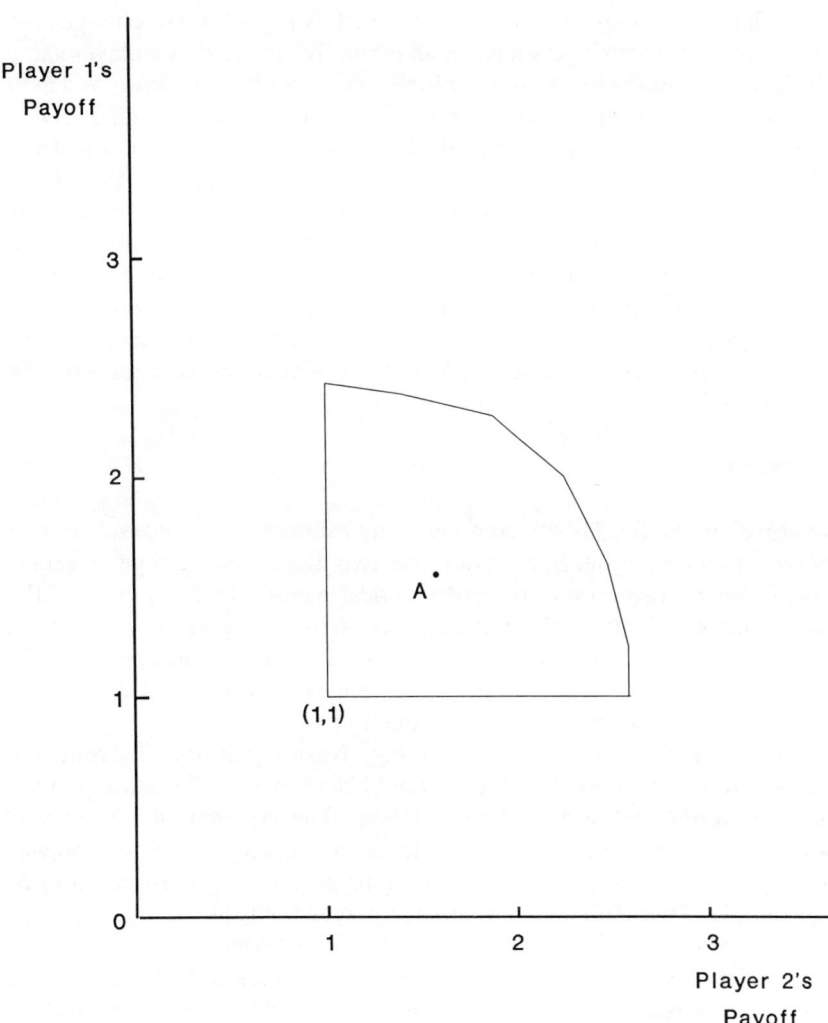

Figure 3–4.

focus for Crawford and Haller is to find how quickly it takes two players to coordinate on an equilibrium. For example, how quickly they can achieve a (1, 1) payoff in Game 1. I then present Aumann and Sorin (1989). They find sufficient conditions on games and beliefs of players so that two players who are playing a game having Pareto ranked equilibria as in Game 2 will always implement the best equilibrium. I conclude the analysis of the purely theoretical papers with a review of Matsui (1989). He enlarges the information sets that players have in an infinitely repeated two-player game and shows that this enlargement will always implement a Pareto efficient outcome in general stage games including the prisoners' dilemma, Game 3. In section 3, I discuss Crawford (1991), which examines a set of coordination experiments conducted by Van Huyck, Battalio, and Beil (1990, 1991) and develops an evolutionary approach to explain their results. A major feature that runs through both the theoretical and empirical pieces of the coordination literature is that coordination is difficult to achieve. The difficulty is demonstrated in the theoretical models by showing the circumstances in which it will always occur. From the empirical side, we find that in practice we typically do not see players coordinating on the best equilibria.

2. Theory

Crawford and Haller (1990) examine many examples to demonstrate their notions of coordination. In each of these, two players play a repeated stage game in which the players receive identical payoffs in every play of the game. The key feature of their analysis is that players do not have a common description of the game. This is the only departure from the assumptions of standard game theoretic analysis. To see what this means, let's examine Game 1 from the introduction.

In Game 1, there are two pure strategy Nash equilibria. Conventional analysis would say that the players would play one of the two equilibria and that would be the end of the analysis. The assumption that players do not have a common description in this game means that there is no way for player 1 to say to player 2, I'm going to play T, so you should play R, for example. Thus, the beliefs that each player holds about what strategy the other player will play are not common knowledge.

If Game 1 were played only a single period, then any strategy that a player uses is optimal, since there is nothing to distinguish which strategy the other player will use. That is, player 1 would think that it is equally likely that player 2 will either play L or R. The focus of Crawford and Haller is to find the optimal strategies for players playing the game. A majority of their paper is devoted to examining different games and ingeniously deriving the optimal strategies that allow players to maximize their joint payoff, and thus coordinate on a pair of strategies as quickly as possible. I think that this analysis is useful when analyzing the pure coordination games without Pareto ranked equilibria. But, one needs to ask what would happen if we perturb the players payoffs a little. For example, what if the players are playing the game shown in figure 3–5? In this game,

COORDINATION IN GAMES: A SURVEY

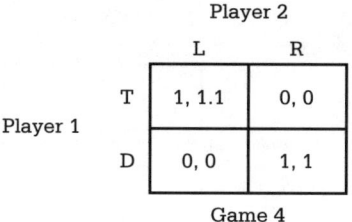

Game 4

Figure 3-5. Game 4

it would seem that (T, L) is a focal equilibrium. This type of game can be better analyzed in the manner of Aumann and Sorin (1989) and Matsui (1989), even though the focus of each of these papers is different.

Aumann and Sorin (1989) stay within the "conventional" two-player repeated game framework when they do their analysis. The goal for Aumann and Sorin is to find sufficient conditions under which the players' payoffs are *always* close to the Pareto dominating payoff pair. That is, when will the players always coordinate their strategies and cooperate on an efficient outcome.

Aumann and Sorin begin their investigation into this question by defining two player games with common interests. A game has common interests if one payoff pair strongly Pareto dominates any other payoff pair. For example, Game 2 is one with common interests, while Game 1, the prisoners' dilemma (Game 3), and the battle of the sexes (Game 5) do not have common interests.[1] Aumann and Sorin would say that players are cooperating in Game 2 if their average per period payoff from playing game 2 were always close to 2. To see why common interests appear to be a necessary condition to support cooperation take Game 5. There are two Pareto efficient equilibria in this game, (T, L) and (D, R). Even though Game 5 does have equilibria which induce Pareto efficient payoffs, there is no guarantee that players will play one of these equilibria. This is because there is also another equilibrium, which has player 1 (2) playing top (right) with probability 2/3. Thus, in a game without common interests players may not agree on what equilibrium to play and thus may play an inefficient equilibrium (see figure 3–6).

Let us now examine why the efficient outcome is not always the one that will be predicted by standard game theoretic arguments even in a game with common interests as in Game 2. In this game, the strategies down and right form a perfectly good strict Nash equilibrium in a one shot game.[2] Thus, it is clear that the game must be repeated. Unfortunately,

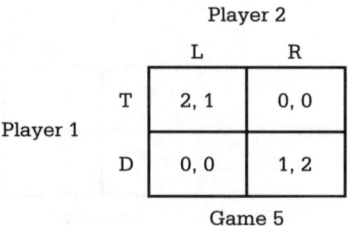

Figure 3-6. Game 5

repetition of this game, or Games 3 and 5, is not sufficient to ensure cooperation. In fact by the folk theorem, the (1, 1) equilibrium payoff pair of Game 2 can be supported in an infinitely repeated game, a supergame, as part of a subgame perfect equilibrium. Thus, even though it is clear that both players would prefer (2, 2), there is no justification for them to focus on the strategies that achieve (2, 2) using the typical refinements. For example, strategies that the row player choose D and column player choose R in all subgames forms a subgame perfect equilibrium.

Thus, the question becomes: What other conditions are necessary in order to guarantee a cooperative outcome? Aumann and Sorin's answer relies on perturbing the game by relying on perturbations from completely rational players. There appears to be no other way to get out of a (D, R) equilibrium in Game 2 and stay within the conventional framework. If the row player will always play D, then it is a best response for the column player to always play R and vice versa even in a supergame. The surprising result is that in order to get cooperation in a game with common interests not only must the game be repeated and perturbed, but the perturbation must take a very particular form. The perturbation deals with the probability that a player may be a finite memory automaton. In particular, all 0 memory automata supergame strategies must be included in the perturbation. Finally, there must be a pure supergame strategy equilibrium of the original game.

I will present the flavor of the proof to see why this particular kind of perturbation is needed to guarantee cooperation. What drives the proof is that, if the players are not playing the Pareto efficient strategy pair, then each player has an incentive to defect to a strategy that mimics the 0 memory perturbation that implements the Pareto efficient outcome. By each player taking this defection, and given that there is no infinite memory automata that might always punish a player for choosing such a defection, the players can move from a bad to a good equilibrium.

COORDINATION IN GAMES: A SURVEY

		Player 2		
		L	M	R
	T	3, 3	1, 1	0, 0
Player 1	M	1, 1	2, 2	0, 0
	D	0, 0	0, 0	0, 0

Game 6

Figure 3–7. Game 6

To see why we need to restrict the perturbation to finite automata, lets look at the following example. Take Game 6, which (figure 3–7) is from Aumann and Sorin, to show that we need bounded recall automata. Suppose that a perturbation from a rational player includes a positive probability that an automaton may have infinite memory. Suppose that the perturbation has an automaton for player 1 (2) playing top (left) if player 2 (1) *does not* play left (top). If player 2 (1) ever does play left (top) then the automaton plays bottom (right) from then on. What we want to now argue is that (M, M) in every stage is an equilibrium. Suppose that player 1 is a rational player and is contemplating a defection to top in period 1 in order to persuade player 2 to play left next period from then on. If there is a sufficiently high probability that player 1 is the "spiteful" automaton, then player 2 will not follow in period 2 because she fears that she will get a 0 in all periods after period 2. Given that player 2 will not follow player 1, a rational player 1 will never defect in period 1 since he lowers his payoff by 1.

Matsui (1989), like Aumann and Sorin (1989), wants to find out when economic agents will choose Pareto efficient outcomes in a two player supergame. As was noted by Aumann and Sorin, infinite by repeated play of a stage game in no way guarantees that an outcome will be Pareto efficient. All that the folk theorem allows us to say is that there is a possibility that the outcome may be Pareto efficient. Matsui's approach is quite different from Aumann and Sorin. Matsui notes that an underlying assumption of the folk theorem is that players can only see the actions taken by others and not their supergame strategies. That is, the players do not see "the book" written at the beginning of the game by each of the other players prescribing what action to take after any conceivable history of the game. Thus, they can only condition their own supergame strategies and actions on only the actions actually taken by their opponents and not their strategies. Matsui studies only two player games. In his main model, Matsui "perturbs" the traditional analysis by giving one of the players a small chance to see the other player's strategy (book), and to revise his own strategy before play actually begins. Matsui shows that if this chance of "espionage" is sufficiently small along with the fact that the cost of revising a strategy is small but positive, then all subgame perfect payoffs are Pareto efficient. That is, information leakage leads to Pareto efficient outcomes. Thus, Matsui does not need to restrict the class of games he analyzes nor does he deviate from completely rational players to obtain his results as was the case in Aumann and Sorin.

The timing of the revision game that is studied by Matsui is the following: first, completely rational players choose their supergame strategies for the play of an infinitely repeated stage game. Next, with probability ε chosen by nature player 2 sees the strategy chosen by player 1. Player 1 does not know whether player 2 has become knowledgeable of her strategy. Player 2 can then choose to change his strategy without regard to the strategy that he originally chose. The players then play an infinitely repeated stage game whose payoffs are determined by these strategies.

It is assumed for most of the paper that player 2's preferences regarding the revision of his strategy are lexicographic. That is, if player 2 can strictly improve his payoff by revising his strategy, then he would prefer to revise it; otherwise, he would prefer not to revise it.

To see how Matsui gets his result let us examine why (D, R) is not an equilibrium in the repeated prisoners' dilemma, Game 3. Player 1 can improve his payoff from this putative equilibrium by adopting the strategy play D in the first stage and in stages 2 and on play T if player 2 has always played L, otherwise play D. With probability ε player 2 sees player 1's strategy. If 2 sees 1's strategy it is optimal to change his strategy to, for example, cooperate the first two periods, play L, and cooperate from then on if player 1 cooperates in all periods after period 1, otherwise choose R. By this deviation both players get 2 per period after period 1 if the revision stage is reached. This deviation pays for both, since 2ε, the approximate per period payoff for each of the players, is greater than 0. What has happened is that player 2 by his choice of strategy has signalled to player 1 that he has seen her strategy and that he will cooperate in the future if player 1 cooperates after period 1. The key to the theorem is that player 2 will act as if he saw player 1's strategy. Player 1 will know that in equilibrium player 2 will mimic his revision behavior so that 1 will choose a strategy that implements a Pareto efficient outcome. This result has some of the same flavor of Aumann and Sorin. Recall that in Aumann and Sorin, a player wants to mimic an automaton that will always play strategies that implement the Pareto dominating payoff pair.

Matsui goes on to show that any Pareto efficient outcome can be supported in the revision game. He also proposes two extensions of the model. First, he allows for both players to have the ability to see each other's strategy. That is, the timing of the game for player 1 is now the same as that for player 2. Matsui shows that if each player has an independently drawn ε probability of seeing the other's strategy and ε is sufficiently small, then an outcome of the revision game is Pareto efficient. He does not show that an equilibrium always exists in the two sided espionage game. The second extension is to allow for a small revision cost, instead of player 2 having lexicographic preferences, in the one sided revision game. He then argues that if player 2 has this type of preferences, then the payoff pair is Pareto efficient.

Matsui points out that in reality we do not see players choose strategies at the beginning of a supergame that states every action that a player will choose given any history of the game. But keeping the same framework, one would like to know the answers to questions such as: What happens to the set of equilibria if the players can influence the probability that another player may learn their strategy? A player might then influence which Pareto efficient outcome will be played by acting as a Stackelberg leader. Another question is: What if the player can only see a part of the other player's strategy?

A feature that has not been investigated in a great deal in these models is the use of cheap talk. Cheap talk are messages (signals) sent by one or more players to the other players in the game, where the messages are not directly payoff relevant. That is, they are not binding and they do not enter the payoff function; however, they may influence the equilibrium of the game by changing players' beliefs concerning other players' strategies. These type of signals should be differentiated from the costly signals in models such as Spence (1973). Rabin (1990) and Zapater (1992) have done work on cheap talk models that allow players to coordinate on Pareto efficient equilibria.

3. Experimental Work

Crawford (1991) analyzes a set of coordination experiments conducted by Van Huyck, Battalio, and Beil (1990, 1991). The experiments each have a set of Pareto ranked strict Nash equilibria. I will describe a subset of the treatments to concentrate on Crawford's reinterpretation of these experiments. All the experiments have a group of subjects simultaneously choosing effort levels in a repeated stage game. Each player's effort choice comes from the same finite set of choices. No communication is allowed between the players during the game. One major difference between the experiments is the payoff function of a player. In the "minimum" experiments, the payoff functions in each period, is $\Pi_i = c + ae_{min} - be_i$, where c is a positive constant, e_{min} is the minimum effort level in the group, e_i is the effort level by player i, and $a > b > 0$. In each stage game, a strict Nash equilibrium exists at every effort level where each player chooses the same level of effort. But, as in Game 2, the equilibria can be Pareto ranked with the higher effort levels resulting in all players receiving higher payoffs. Thus, we should expect that if players could coordinate they would each choose the highest allowable effort level. The experiments within the minimum set differ by the groups either including all the experiments participants, "large" groups with 14–16 participants or having pairs of players randomly matched in every period.

The results for the large group experiments have the following characteristics: each effort level was chosen initially by some agent with a majority choosing effort levels above the median of the set. As the experiment continued the effort choices by all subjects converged to the minimum effort in most treatments. In the random pairing treatments, the effort choices "drifted over time with no discernable trend" (Crawford p. 33).

In the other set of experiments, player i's payoff function is $\Pi_i = d + ae_{med} - b[e_{med} - e_i]^2$, where e_{med} is the median effort by the group and $a > b$. Thus, as in the minimum experiments there are many pure strategy strict Nash equilibria in each stage game, with the Pareto dominating equilibria at the highest effort level. There were 9 subjects in each group. In most of the treatments of this experiment, the subjects converged quickly to the first period median. Thus, while in the large group minimum experiments the groups generally converged to the lowest level of effort that was allowable, the effort levels that a group converged to in the median experiments was largely determined by the first period's play of the game.

Traditional game theory provides us no guidance for why one particular equilibrium is played over another. Furthermore, it is quite striking to see that by going from the minimum to the median experiments, history goes from not mattering at all in a large group setting to the first period outcome essentially dictating how the rest of the game should be played. Why do we see the worst equilibrium in the minimum model, while we see history dependent equilibria in the median model? Crawford first proposes a purely evolutionary explanation for these result and then elaborates on this to allow for players to draw inferences to provide a fuller explanation.

The basic idea behind evolutionary game theory, see Maynard Smith (1982), is to have a large set of identical players (animals) each use some exogenously given pure strategy. Each player's payoff is determined by how well he does relative to the population average. Given that the population is large an individual player does not believe that his play effects the population frequency. The dynamics of population frequency have the property that player (strategy) types with higher payoff increase there frequency over time. Thus, higher payoff strategies will be present over time. In general, attention in these games is restricted evolutionary stable stationary strategies (ESS). A strategy is an ESS in a large population if it performs better than any mutant strategy for all frequencies of the mutant strategy less than ε, where ε is a small number. Thus, attention is restricted to a local stability condition.

Following Crawford, we will now reanalyze the experiments where there are only two possible effort levels e_L and e_H, with $e_H > e_L$. Furthermore, let $a = 2$ and $b = 1$. To see why the e_L is the only ESS in the large minimum game, while both e_L and e_H are both ESS in the small minimum, random pairing, games examine Figures 3–8 and 3–9. In figure 3–8, the payoffs are on the vertical axis, while the frequency of e_L in the population is on the horizontal axis. Notice that for *any* fraction of players using strategy e_L, the payoff using e_H has a player receiving a strictly lower payoff than a player using e_L; she gets 1 instead of 0. That is, there is a downward discontinuity in a player's payoff function if he uses e_H. On the other hand, Figure 3–9 represents a player's payoffs in the random pairing model. In this game, there are two stable ESS, both at e_H and e_L. This is because there is no downward discontinuous jump in a player's payoff if she chooses effort level 2 and with a small probability she meets a player who chooses an effort level of 1. Thus, as in the experiments with random pairing where the effort levels did not necessarily collapse to the lowest level, evolutionary theory would not necessarily imply that it does in the random pairing model. In the median experiments, any strategy where all players use the same effort level is an ESS. This is because no small invasion can change the median. Thus, other ESS besides e_L can be supported in the median games. Of course, evolutionary theory at this level does not tell us why history would matter. Crawford enriches the model to show that allowing players to adapt their behavior can provide a role for history in an evolutionary framework.

He does this by allowing players' strategy choices to be voluntary in an evolutionary context, but the beliefs about what other players are doing

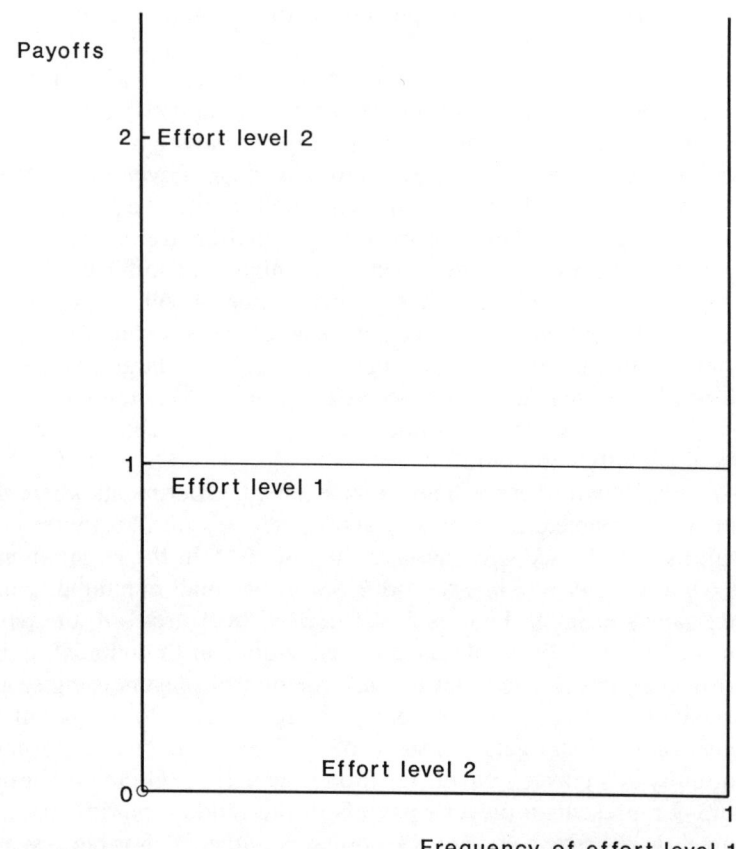

Figure 3–8.

are involuntary. That is, differences in players' beliefs can act as mutations to drive the dynamics. There are two types of differences that are highlighted by Crawford. First, players may have differences in their initial beliefs about what equilibrium effort level other players will focus on. Second, players experience small independent idiosyncratic zero mean shocks, perturbations of their beliefs, regarding how they interpret the history of the game. Crawford argues that if players differ only by their initial beliefs, that most reasonable adjustment processes would have the players coordinate on the initial minimum or median.

The consequence of having only this type of perturbation would explain

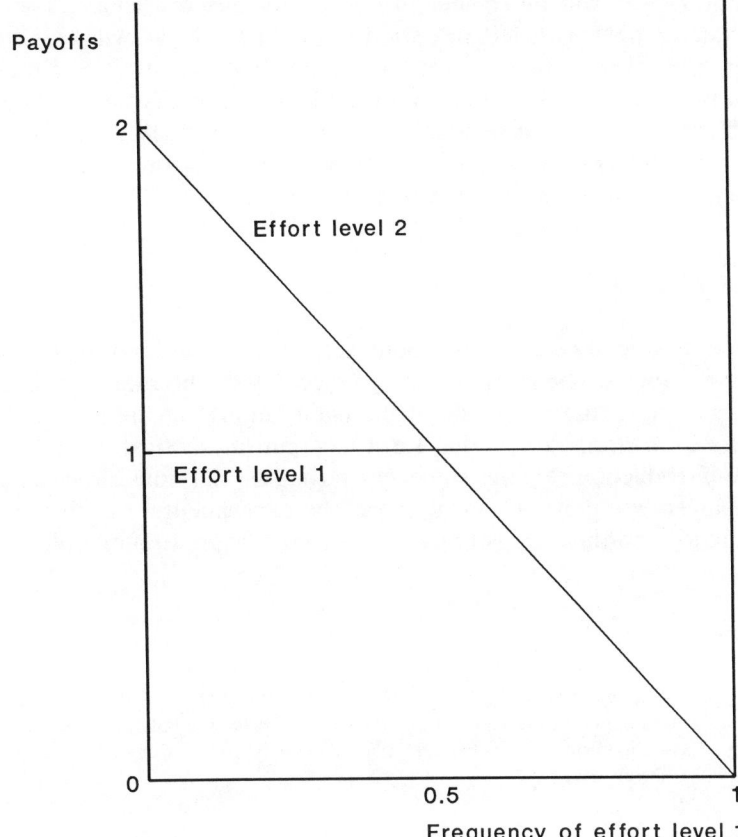

Figure 3-9.

the median experiments, but not the large minimum experiments. Crawford then adds the second type of perturbation to his model. The players are most likely to expect next period's minimum or median to equal last period's, but with a small probability the player thinks that next period's minimum or median will be above or below last period's outcome. Furthermore, the idiosyncratic error terms become less important as the game proceeds. To see how the addition of this perturbation can change the predictions of the equilibrium for the minimum and median games, we should do the following thought experiments. Take the minimum game. For any given minimum in period t, some player will think that the minimum in period $t + 1$ will be smaller in $t + 1$, and thus will choose the effort associated with her beliefs in period $t + 1$. In $t + 1$, we will then have a lower minimum than in t, this process will proceed until either the perturbations go to 0, or as in most of the experiments the lowest effort is achieved. In the median games, the perturbations do not make, on average, large changes on the group median. Thus, the group's initial median is very likely to be the median that they converge to.

4. Conclusion

In the introduction some ideas about coordination were introduced. I then reviewed some of the more prominent pieces of the literature dealing with coordination. I think that one of the main findings of the theoretical literature on coordination is that for it to occur in any of the senses stated in the introduction requires some very strong assumptions about the game that players are playing. Furthermore, the experimental work shows that in practice coordination is quite difficult for players to accomplish.

Notes

1. It is not necessary for the players' payoffs to be identical for a game to have common interests. For example, the row player's payoffs can be twice the column player's payoff.
2. A Nash equilibrium is strict if each player's best response is unique given the other player's equilibrium strategy.

References

Aumann, R., and S. Sorin. 1989. "Cooperation and Bounded Recall." *Games and Economic Behavior* 1: 5–39.
Crawford, V. 1982. "A Theory of Disagreement in Bargaining." *Econometrica* 50: 607–637.
Crawford, V. 1991. "An 'Evolutionary' Interpretation of Van Huyck, Battailo, and Beil's Experimental Results on Coordination." *Games and Economic Behavior* 3: 25–59.
Crawford, V., and H. Haller. 1990. "Learning How to Cooperate: Optimal Play in Repeated Coordination Games." *Econometrica* 58: 571–596.
Friedman, J. 1971. "A Non-cooperative Equilibrium for Supergames." *Review of Economic Studies* 38: 1–12.
Fudenberg, D., and E. Maskin. 1986. "The Folk Theorem in Repeated Games with Discounting or with Incomplete Information." *Econometrica* 54: 533–556.

Matsui, A. 1989. "Information Leakage Forces Cooperation." *Games and Economic Behavior* 1: 94–115.

Maynard Smith, J. 1982. *Evolution and the Theory of Games*. New York: Cambridge University Press.

Rabin, M. 1990. "Communication Between Rational Agents." *Journal of Economic Theory* 51: 144–170.

Schelling, T. 1960. *The Strategy of Conflict*. Cambridge, MA: Harvard University Press.

Van Huyck, J., R. Battalio, and R. Beil. 1990. "Tacit Coordination Games, Strategic Uncertainty, and Coordination Failure." *American Economic Review* 80: 234–248.

Van Huyck, J., R. Battalio, and R. Beil. 1991. "Strategic Uncertainty, Equilibrium Selection, and coordination Failure in Average Opinion Games." *Quarterly Journal of Economics* 885–910.

Zapater, I. 1992. "Generalized Communication Between Agents." Brown University Working Paper.

II GENERAL ISSUES IN COORDINATION

4 INCORPORATING BEHAVIORAL ASSUMPTIONS INTO GAME THEORY

Matthew Rabin*

1. Introduction

The standard approach to making predictions in noncooperative game theory is to invoke internal consistency: Behavior is only ruled out when we can argue that if players came to believe in the behavior, then at least one player would wish to deviate. Such internal-consistency arguments clearly underlie the solution concepts Nash equilibrium and rationalizability, and arguably underlie most prevalent game-theoretic solution concepts.[1] By now, most of us perceive an apparent shortcoming of this approach to noncooperative game theory: Using even the most strained arguments about what rationality implies, analyses of many games do not yield sharp predictions.

---

* This chapter is based in part on chapters 1–3 of the dissertation (Rabin [1989]) completed at MIT. The author thanks Eddie Dekel-Tabak, Joe Farrell, Oliver Hart, Joel Sobel, and especially Drew Fudenberg for their contributions, as well as the National Science Foundation and the Alfred P. Sloan Foundation for financial support during that period. He also thanks Jim Friedman for very useful suggestions on the current draft.

I believe that game theorists must make their peace with these limits. We ought incorporate "behavioral assumptions"–assumptions about how play might be limited not solely by the structure of a game, but also by players' shared beliefs about what is likely behavior. Thus, we might rule out some internally consistent behavior because we do not think that the players will come to believe in such behavior. In this paper, I develop a way of formulating solution concepts that combine behavioral assumptions with standard views of internal consistency incorporated into both Nash equilibrium and Bernheim's (1984) and Pearce's (1984) notion of *rationalizability*.

Recently, the use of behavioral assumptions has become widely accepted in the literature on "cheap talk."[2] When communication is not costly, arguments from the point of view of internal consistency can never guarantee communication. This literature studies how we can make stronger predictions about outcomes in many situations by making the behavioral assumption that people expect truthful communication. Consider figure 4–1, which represents a simple, simultaneous-move coordination game preceded by an opportunity for player 1 to make a suggestion on the play of the game.

Most of us would have little trouble believing that the outcome here will be that player 1 suggests either (U, L) or (D, R), and the players play the equilibrium that player 1 suggests.[3] The incentives in this game are such that honest communication by player 1 about his intentions is very credible, and the desirable coordination is likely to be achieved. Yet no solution concept from rationalizability to Nash equilibrium to strategic stability guarantees any communication here. Because meaningful communication is not guaranteed by any of these solution concepts, it is in turn not guaranteed that the players will be able to coordinate on either of these equilibria.

To predict that communication and coordination would occur in Figure 4–1, we must invoke behavioral assumptions. That is, we must assume *a priori* that the players will have the propensity to communicate honestly. The perspective of this paper is what we may more generally be able to invoke simple behavioral regularities to establish which outcomes in an economic situation are likely to occur. While my purpose is ultimately to incorporate empirically valid behavioral assumptions, my focus in this paper is on the more fundamental question of how one adds behavioral assumptions without abandoning standard assumptions about rationality. Just as we have developed a language to express coherent structural assumptions (such as perfect recall and common priors), so too I believe that we can do the same for *behavioral* assumptions.

INCORPORATING BEHAVIORAL ASSUMPTIONS INTO GAME THEORY 71

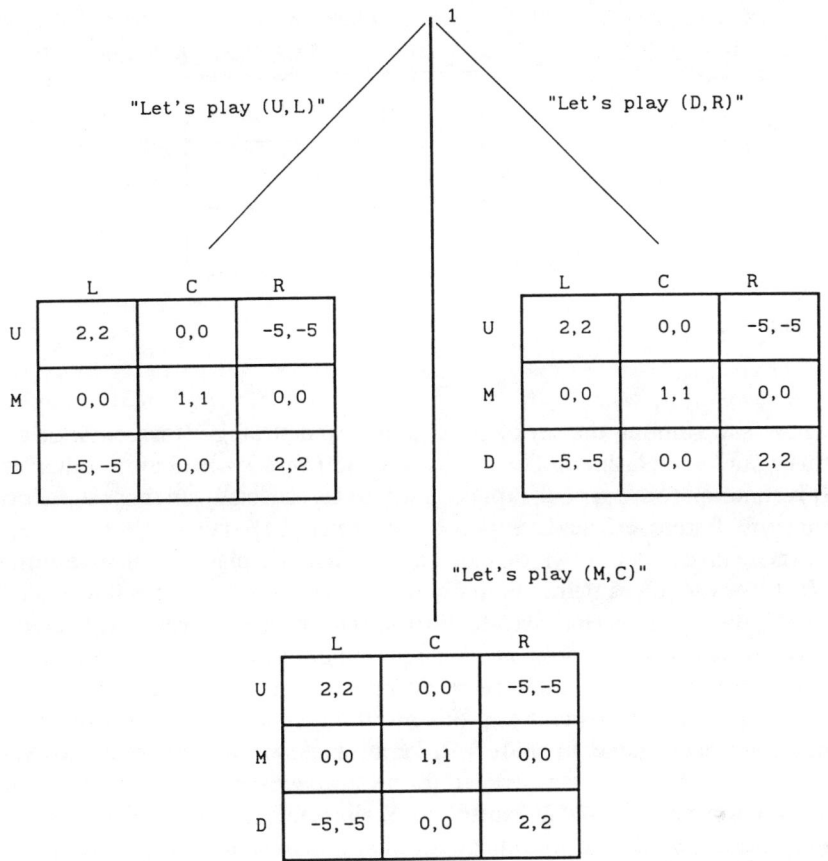

Figure 4–1.

My approach is simple. I assume that, while players enter into a strategic situation with a set of *possible* strategies, they contemplate their choices in terms of some set of *focal* strategies that they imagine are the only strategies that are likely to be played. Once they so imagine, they play the game as if this focal set of strategies is the real game. In figure 4–1, for instance, we could postulate that only strategies in which player 2 will behave according to player 1's suggestion will be focal. Then, if player 1 chooses his strategy believing that player 2 is behaving this way, he will surely propose one of the efficient equilibria.

To illustrate this approach in a noncommunicational example, consider

Figure 4–2.

		L	C	R
	U	2, 1	0, 0	−5, −5
Player 1 M		0, 0	1, 2	−5, −5
	D	−5, −5	−5, −5	−2, −2

Player 2 appears as the column header above.

figure 4–2. Among the set of pure- and mixed-strategy Nash equilibria in this game, two Nash equilibria—(U, L) and (M, C)—are Pareto-efficient.[4] It is common to suppose that players will focus especially on those equilibria that are Pareto efficient. Suppose we argue that such a focus is likely in this game, but that we do not believe that the players perceive either (U, L) or (M, C) as being uniquely focal. It is therefore natural to believe that player 1 will select a strategy from $\{U, M\}$, and player 2 will select a strategy from $\{L, C\}$. Formally, we might hypothesize that the players would play this game "as if" there were no option to play either D or R.

My approach to formalizing this intuition is to invoke rationalizability in the restricted game. In figure 4–2, for instance, we can assume that players will act as if the game is restricted to the strategies $\{U, M\}$ for player 1 and $\{L, C\}$ for player 4–2. Because rationalizability is unrestrictive in this hypothetical game, we would assume that any outcome in $\{U, M\} \times \{L, C\}$ is possible.

Are all solution concepts equivalent to the application of rationalizability to a hypothetically restricted game sensible? The answer is no: To avoid wholly arbitrary beliefs and outcomes, we must in many games impose some further restrictions. Figure 4–3 helps illustrate the need for such restrictions. As in Figure 4–2, the two Pareto-efficient Nash equilibria in figure 4–3 are (U, L) and (M, C). Suppose that, as in figure 4–2, we think it natural for the players to focus on the hypothetical game in which they cannot play D or R.

Applying rationalizability to this hypothetical game, we would predict that any combination of U and M might be played by player 1, and any combination of L and C might be played by player 2. For instance, player 1 might believe that player 2 is playing L with probability .25, and C with probability .75. In the hypothetical game, player 1 would respond to such

		Player 2		
		L	C	R
	U	30, 10	0, 0	-5, 9
Player 1	M	0, 0	10, 30	-5, 9
	D	9, -5	9, -5	-2, -2

Figure 4–3.

beliefs by playing U. Yet in the *real* game, where player 1 is allowed to choose the strategy D, he would in fact choose D in response to these beliefs. Using similar arguments for player 2, we realize that even if the players will "tend to focus" their beliefs and behavior on (U, L) and (M, C)—but do not know which of these equilibria is most reasonable—we cannot rule out the outcome (D, R).

Thus, to check that a behavioral theory is consistent with rationality, I impose the conditions 1) that it be equivalent to rationalizability in some hypothetically restricted game, but also 2) that it be consistent with rationality in the actual game. In Sections 2 and 3, I develop the concept of *Consistent Behavioral Theories* to formalize these notions of consistency.[5] The set of predictions $\{U, M\} \times \{L, C\}$ constitute a consistent behavioral theory in figure 2, but not in figure 4–3.

Consistent behavioral theories are behavioral analogs of rationalizability, as defined by Bernheim (1984) and Pearce (1984). The question arises: What are the behavioral analogs to Nash equilibrium, perfect equilibrium, etc.? This is harder to say, because many equilibrium concepts have been motivated and interpreted by different people in different ways. Nonetheless, I propose in section 4 the definition *Behavioral Equilibrium Theory* as the behavioral analog to Nash equilibrium.

In section 5, I discuss a means of developing consistent behavioral theories using an algorithm which, beginning with a particular set of focal strategies, iteratively expands the set of strategies by adding in all strategies that are rational responses to beliefs by the players. In section 6, I use this method to develop a consistent behavioral theory that incorporates the idea that players tend to focus their beliefs and behavior on Pareto-efficient Nash equilibria. I also discuss more generally the role of behavioral assumptions in helping players coordinate on efficient outcomes.

I conclude in section 7 with a discussion of possible extensions and refinement of the framework developed in this paper, highlighting the fact that this paper focuses solely on behavioral analogs to rationalizability and Nash equilibrium, rather than on the well-known refinements of these concepts.

2. Preliminaries

Consider a two-player, normal-form game G, where players 1 and 2 have the set of pure strategies S_1 and S_2 with n_1 and n_2 elements, respectively.[6] For each ordered pair of strategies, $(s_1, s_2) \in (S_1, S_2)$, let $(U_1(s_1, s_2), U_2(s_1, s_2))$ be the payoffs for each of the two players. Let Σ_1 and Σ_2 be the sets of mixed strategies for players 1 and 2. The expected payoffs over mixed strategies $(\sigma_1, \sigma_2) \in (\Sigma_1, \Sigma_2)$ can be represented by the functions $(v^1(\sigma_1, \sigma_2), v^2(\sigma_1, \sigma_2))$, derived from the utility functions U_1 and U_2.

The approach I develop will incorporate beliefs more explicitly than is conventional; because of this, I now develop some nonstandard notation and terminology. Define a *plan* θ_i for each player i as a pair (σ, μ), where $\sigma \in \Sigma_i$ is a $1 \times n_i$ vector representing a mixed strategy for player i, and $\mu \in \Sigma_j$ is a $1 \times n_j$ vector representing player i's beliefs about the strategy employed by player j. A player's plan includes both the strategy he chooses, and his beliefs about the other player's strategy. Let Θ_i be the set of all possible plans for player i; Θ_i can be represented as a set of points in $\mathbf{R}^{n1+n2}$, representing both strategies and beliefs as likelihood distributions over S_1 and S_2. A *prediction* for player i, p_i, is a likelihood distribution over the set. Of all possible elements in Θ_i. Let $\mathcal{P}_i$ be the set of all possible predictions over Θ_i. For every $p_i \in \mathcal{P}_i$, and for every $\theta \in \Theta_i$, define $p_i(\theta)$ as the likelihood that p_i places on θ. Whereas $(\mathcal{P}_1, \mathcal{P}_2)$ are sets of likelihood distributions over plans, it will frequently be useful to separately examine the implied distribution over strategies. For a given $p_i \in \mathcal{P}_i$, define p_i^s as the probability distribution over player i's strategies $s_i \in S_i$ derived from p_i; $p_i^s = \int_{(\sigma,\mu)\in\Theta i} p_i(\sigma,\mu) \cdot \sigma$.

I say that a pair of sets $(P_1, P_2) \subseteq (\mathcal{P}_1, \mathcal{P}_2)$ of predictions is a *theory*. Because predictions over the strategies of the players will often be of interest, I define the sets (P_1^s, P_2^s), where $P_i^s = \{p_i^s | p_i \in p_i\}$. These are the more traditional data of analysis.

A central tenet of game theory is that players behave rationally: For all plans $\theta_i = (\sigma, \mu)$ given positive weight by some prediction, the strategy σ ought to be optimal given the beliefs μ. Formally, σ^* is a *best response* for player i to the beliefs μ if $\sigma^* \in \mathrm{argmax}_{\sigma \in \Sigma_i} v^i(\sigma, \mu)$. Define the sets of predictions $(\mathcal{P}_1^*, \mathcal{P}_2^*) \subseteq (P_1, P_2)$ as the set of predictions that put positive

INCORPORATING BEHAVIORAL ASSUMPTIONS INTO GAME THEORY   75

weight only on plans for which the strategies are best responses to beliefs. $(\mathcal{P}_1^*, \mathcal{P}_2^*)$ are the sets of predictions consistent with players being rational.

To define those predictions that are consistent with *common knowledge* of rationality, I shall construct the concept that is the analog within this framework of the solution concept rationalizability developed by Bernheim (1984) and Pearce (1984). The construction here parallels one of the two equivalent constructions of rationalizability in Pearce (1984). I begin with the definition of sets that have the *best-response property*:

Definition 1. The sets of predictions $(A_1, A_2) \subseteq (\mathcal{P}_1^*, \mathcal{P}_2^*)$ have the *best-response property* iff for $i = 1, 2$, for $j \neq i$, for all $\theta_i = (\sigma, \mu) \in \Theta_i$ with $p(\theta_i) > 0$ for some $p \in A_i$, there exists a $p_j \in A_j$ such that $\sigma = p_j^s$.

Sets of predictions for players have the best-response property if all beliefs the theory allows for one player correspond to some behavior that the theory allows for the other player. If a strategy belongs to *any* sets with the best-response property, then it is consistent with common knowledge of rationality. This is because any strategy in A_1 is a utility-maximizing response for player 1 to some beliefs over the plans in A_2 employed by player 2, each of which are optimal responses to some beliefs over the plans in A_1 employed by player 1, each of which are optimal responses to some beliefs over the strategies in A_2 employed by player 2, etc. So long as two sets meet the best-response property, each prediction for each player corresponds to some such infinite sequence of rational strategy choices.

From Definition 1, it is straightforward to characterize the set of all predictions consistent with common knowledge of rationality.

Definition 2. The set of *Rationalizable Predictions* is (R_1, R_2) in which $p_i \in R_i$ if and only if there exists some (A_1, A_2) with the best-response property such that $p_i \in A_i$.

Pearce illustrates that this definition of rationalizability is equivalent to the process of iterated deletion of dominated strategies.

Clearly, the set of rationalizable strategies has the best-response property. It can also be shown that the set of rationalizable predictions has several other features. Because these other features are useful for some of the definitions below, I now present them:

Definition 3. The sets of predictions $(A_1, A_2) \subseteq (\mathcal{P}_1^*, \mathcal{P}_2^*)$ have the *covering property* if for $i = 1, 2$, for $j \neq i$, for each $p_j \in A_j$, there exists some $p_i \in A_i$ such that for all (σ, μ) for which $p_i(\sigma, \mu) > 0$, $\mu = p_j^s$.

Sets of predictions have the covering property if, for any behavior that the theory allows for one player, the theory allows that the other player believes with certainty in that behavior. This is sort of the converse of the best-response property.

Definition 4. (A_1, A_2) is said to be *convex* if for all $p_i, p_i^* \in A_i$, and for all $\lambda \in [0, 1]$, then $p_i^{**} \in A_i$, where for all $\theta_i \in \Theta_i$, $p_i^{**}(\theta_i) = \lambda \cdot p_i(\theta_i) + (1 - \lambda) \cdot p_i^*(\theta_i)$.

By imposing convexity of predictions for a player, we essentially impose convexity on the sets of permissible beliefs for both the game theorist and the other players.

Definition 5. For the sets (A_1, A_2), say that σ_i is a component strategy for player i if there exists a prediction $p_i^* \in A_i$ and $\mu \in A_j^s$ such that $p_i^*(\sigma_i, \mu) > 0$.

(A_1, A_2) is said to *be regular* if, for $i = 1, 2$, then $p_i \in A_i$ if and only if it puts positive weight on plans (σ, μ) such that

1. σ is an optimal response to beliefs μ,
2. $\mu \in A_j^s$, and
3. σ is a component strategy for player i.

This essentially says that a strategy is employed against beliefs for which it is optimal if and only if it is employed against all beliefs in the theory for which it is optimal. As the name suggests, this will used as a convenient regularity condition.[7]

It is relatively straightforward—and useful for later purposes—to show that the set of rationalizable strategies have all of these properties:

Lemma 1. The set of rationalizable predictions is convex and regular, and have the best-response and covering properties.[8]

3. Consistent Behavioral Theories

In this section, I present the notion of *Consistent Behavioral Theories (CBTs)*, which formalizes the criteria for behavioral theories outlined in section 1.

First, consider the implications of the players restricting their beliefs

INCORPORATING BEHAVIORAL ASSUMPTIONS INTO GAME THEORY   77

and behavior to some closed, convex subsets of mixed strategies $Q \equiv (Q_1, Q_2)$, $Q_k \subseteq \Sigma_k$ for $k = 1, 2$. I restrict attention to convex sets because that is the natural implication of people's belief-formation—as in considering real games, we allow players to have any conceivable mixes of beliefs.

Let $\Gamma(G, Q)$ be the hypothetical game with strategy sets Q and payoffs $V^Q(\Sigma)$ such that, for all $s \in Q$, $V_k^Q(s) \equiv V_k(s)$. This game is well-defined, with strategy spaces that are subsets of the strategy spaces in G, and payoffs that are identical to G. We can therefore define any solution concept in this hypothetical game. A natural starting place is the solution concept with the most intimate connection to the rationality assumption—rationalizability. For a game G and chosen convex subsets of strategies, Q, let $(R_1(G, Q), R_2(G, Q))$ be the set of rationalizable strategies in the game $\Gamma(G, Q)$. If we let $\mathcal{P}_i^*(G, Q)$ be the set of rational predictions in the game $\Gamma(G, Q)$, then $R_i(G, Q)$ is by construction a subset of $\mathcal{P}_i^*(G, Q)$. Importantly, however, $R_i(G, Q)$ is *not* necessarily a subset of $\mathcal{P}_i$; this is because we do not know that best responses for player i when restricted to strategies in Q_i are also optimal when he can employ any strategy in Σ_i.

I say that any sets of strategies that correspond to rationalizability in some such hypothetical game is a *behavioral theory*.

Definition 6. The set of predictions (P_1, P_2) constitute a *Behavioral Theory* if there exists $(Q_1, Q_2) \subseteq (\Sigma_1, \Sigma_2)$ such that $(P_1, P_2) = (R_1(G, Q), R_2(G, Q))$.

A behavioral theory is a set of predictions that would be consistent with common knowledge of rationality if the players were restricted to employ strategies only in (Q_1, Q_2). Are all behavioral theories reasonable? It is quite possible they are not, if we consider that the real game does not restrict the players to choose from sets of strategies Q_k, but rather from sets Σ_k.

Figure 4–3 illustrates the problem: the set of rationalizable predictions on the hypothetical game consisting of strategies $\{U, M\} \times \{L, C\}$ are not rational in the actual game. Thus, in such a behavioral theory, not all predictions involve a player responding rationally given his beliefs. The next definition rules out irrational predictions:

Definition 7. A behavioral theory (P_1, P_2) is *a Consistent Behavioral Theory (CBT)* if $(P_1, P_2) \subseteq (\mathcal{P}_1^*, \mathcal{P}_2^*)$.

This definition simply says that the theory thus derived must have each player responding rationally given his beliefs, and given his unrestricted choice of strategies. The idea of a CBT essentially characterizes behavioral theories which can be common knowledge to the players, in the sense that if the players hold any beliefs consistent with the theory, they do not (strictly) prefer to deviate from the theory. In this way, they are similar to rationalizability: if players believe with common knowledge in the set of rationalizable strategies, then no rational player would want to deviate to play a nonrationalizable strategy.

From this relatively simple construction, much can be implied about these sets. Indeed, it is straightforward to show that every CBT has the same properties as the set of rationalizable strategies, as outlined in section 2. It turns out that the converse is also true: if a pair of sets $(P_1, P_2) \subseteq (\mathcal{P}_1^*, \mathcal{P}_2^*)$ have the four properties of rationalizability, then they constitute a CBT.

Theorem 1. (P_1, P_2) is a CBT if and only if it is convex and regular, and has the best-response and covering properties.[9]

Proof. The "only if" direction is clear. In the "if" direction, we need only observe that $(P_1, P_2) = (R_1(G, Q), R_2(G, Q))$ where $(Q_1, Q_2) = (P_1^s, P_2^s)$. Because $(P_1, P_2) \subseteq (\mathcal{P}_1^*, \mathcal{P}_2^*)$, these sets constitute a CBT.

The proof simply observes that, if we create a hypothetical game from the set of strategies implied by a set of predictions with the four properties, then applying rationalizability on this game will yield the set of predictions we started out with. Then, clearly, this set will be a CBT, because the sets of predictions are rational given the real game.

Thus, these four conditions are necessary and sufficient for a set of predictions to be a CBT. In constructing CBTs, we can therefore look for sets of predictions that meet these criteria, rather than applying the two-step, hpypothetical-game formulation.

4. Equilibrium Theories

The previous section outlined a general approach to formulating solution concepts that are behavioral analogs of rationalizability. In this section, I propose a test for whether an equilibrium theory is a behavioral analog to Nash equilibrium.

Suppose that the players believe with common knowledge that they will play strategies from some hypothetical game (Q_1, Q_2) meeting the consistency criterion from the previous section. Then we can predict that they will play strategies from the corresponding CBT. Now suppose that the only further hypothesis we have about behavior is that only Nash equilibria will be played. Then an outcome is plausible if and only if it is Nash equilibrium contained in the CBT. I call such an equilibrium concept that combines Nash equilibrium with some common-knowledge behavioral restrictions *a Behavioral Equilibrium Theory (BET)*.

If an equilibrium solution concept is not a BET, then it must involve a motivation beyond the common-knowledge behavioral restrictions, or beyond the basic equilibrium hypothesis. Suppose that, as I will shortly demonstrate can be the case, some subset of Nash equilibria in a game does not correspond to the set of Nash equilibria for *any* CBT. Then if we invoke such a subset of Nash equilibria as a solution concept, then we must be invoking some restriction on behavior beyond common-knowledge behavioral restrictions and the basic equilibrium hypothesis.

Of course, such restrictions might make sense. There exist internal-consistency arguments for eliminating some Nash equilibria, as well as many dynamic stories that attempt to model in strategic situations the implications of evolution or learning over time.[10] I discuss in the concluding section how one might combine such arguments with behavioral restrictions.

The concept of BET essentially constitutes a specific method of refining Nash equilibrium based *solely* on common-knowledge behavioral assumptions. Namely, we should use the behavioral assumptions to first refine rationalizability, and then use as our equilibrium concept all the Nash equilibria contained in the corresponding CBT.

To illustrate the idea of a BET, consider figure 4–2 again, and consider also the behavioral assumption that players have a tendency to focus on strategies consistent wit Pareto-efficient Nash equilibria. From such a theory, we can consider the CBT that includes all rationalizable predictions in the game excluding beliefs and behavior focused on the strategies D and R.

The set of Nash euilibria consistent with this CBT are the two Pareto-efficient Nash equilibria, *and* the inefficient mixed-strategy Nash equilibrium (2/3U, 1/3M; 1/3L, 2/3C). That is to say, if we incorporate into a common-knowledge behavioral assumption the idea that players tend to focus on Pareto-efficient Nash equilibria, we discover that the players may still play an inefficient Nash equilibrium. Interestingly, it can be shown that *any CBT that contains both the equilibria (U, L) and (M, C) in figure 4–2 also contains the mixed-strategy Nash equilibrium over these strategies*.[11] This means that, in this game, the solution concept "Pareto-efficient Nash equilibrium" is not a BET based on *any* CBT.

Figure 4–3 illustrates the limited powers of behavioral assumptions even more strikingly than figure 4–2. It can be shown that *any* CBT containing the two efficient Nash equilibria also contain *all* Nash equilibria in this game. The assumption that people tend to focus on Pareto-efficient Nash equilibria has no behavioral implication in this game.

I now present a definition and a theorem allowing us to define BETs formally:

Definition 8. For a CBT (P_1, P_2), define its set of *equilibrium predictions* as the set of strategy pairs $P^e \subseteq \Sigma_1 \times \Sigma_2$ such that
$(\sigma_1, \sigma_2) \in P^e$ iff there exists a $(P_1, P_2) \in (P_1, P_2)$, such that
for $i = 1, 2$, $\sigma_i = P_i^s$; and
for $i = 1, 2$, $j \neq i$, for all (σ, μ) such that $p_i(\sigma, \mu) > 0$, $\mu = \sigma_j$.

The following theorem is an immediate corollary to the fact that the set of rationalizable predictions always includes Nash equilibrium:

Theorem 2. If $P = (P_1, P_2)$ is a CBT, then P^e is non-empty.

I can formally define a *a Behavioral Equilibrium Theory*:

Definition 9. A set E is *a Behavioral Equilibrium Theory (BET)* if there exists a CBT (P_1, P_2) such that $E = P^e$.

The above argument shows that the theory "Pareto-efficient Nash equilibrium" is not a BET—in addition to a behavioral assumption, it must involve some theory of coordination beyond the Nash-equilibrium hypothesis.[12] I discuss in section 6 a CBT and BET that make Pareto-efficient Nash equilibria focal in a game.

In all of these examples we could argue that payers are likely to focus on a particular Nash equilibrium. It would then be a consistent behavioral theory to simply predict this equilibrium as the outcome. Indeed, any time we propose as our theory in a game that *a particular* Nash equilibrium will occur, then that pair of predictions meets the criteria of a CBT.

Theorem 3. For any Nash equilibrium (σ_1, σ_2), the theory $(\{p_1\}, \{p_2\})$, where $p_1(\sigma_1, \sigma_2) = 1$ and $p_2(\sigma_2, \sigma_1) = 1$, is a CBT and a BET.

If the players know which equilibrium will obtain, then predicting this equilibrium can be a sound behavioral theory—it is both a CBT and a BET, Thus, every Nash equilibrium in every game is contained in at least two BETs—the prediction by itself, and the entire set of Nash equilibria.

The restrictiveness of the BET approach therefore is that its says which combinations of equilibria can together exclusively and exhaustively constitute a solution concept.

5. Constructing CBTs from Focal Sets

While the formal argument for CBTs essentially involves beginning with a set of strategies, and then reducing this set, much of the intuition in the examples has worked in the opposite direction—we posited a particular focal set of strategies, and expanded the set to include more predictions.

This procedure can be formalized:

Definition 10. Choose any subsets of predictions $(A_1, A_2) \subseteq (\mathcal{P}_1^*, \mathcal{P}_2^*)$. Then let the *Maximal Expansion of (A_1, A_2)* be the set $Z(A_1, A_2) \subseteq (\mathcal{P}_1^*, \mathcal{P}_2^*)$ constructed as follows:
Let $(A_1(0), A_2(0)) \equiv (A_1, A_2)$.
Then, for integers $k > 0$, $i = 1, 2, j \neq i$,
Let $\mathcal{S}_j(k - 1) \equiv$ the convex hull of $A_j^s(k - 1)$.
Let $A_i(k) \equiv \{p_i \in \mathcal{P}_i^* | \forall\ (\sigma, \mu) \text{ such that } p_i(\sigma, \mu) > 0, \mu \in A_j^s(k - 1)\}$.
Then $Z(A_1, A_2) \equiv \text{Lim}_{k \to \infty}(A_1(k), A_2(k))$.

This procedure involves beginning with a set of rational predictions for each player. We then iteratively add all predictions for each player that involve best responses to beliefs that are consistent with the behavior of the other player. We do so until we add in no more predictions for either player.[13] The use of this procedure is indicated by the following theorem:

Theorem 4. If a set of predictions $(A_1, A_2) \subseteq (\mathcal{P}_1^*, \mathcal{P}_2^*)$ has the best-response property, then its Maximal Expansion $Z(A_1, A_2)$ is a CBT.

Proof. The maximal expansion will clearly itself have the best-response property; moreover, by the fact the maximal expansion is constructed as a limit, it will have the other three properties.

Note that the set of predictions for players corresponding to any subset of Nash equilibria constitute a set with the best-response property. This immediately implies that if we apply a maximal expansion to any subset of Nash equilibria, we will end up with a CBT. I apply this method to the set of Pareto-efficient Nash equilibria in the next section.

6. Coordinating on Efficient Outcomes

The examples in this paper have emphasized the role of behavioral assumptions in helping players coordinate on efficient outcomes.[14] In figure 4-1, I argued that the existence of a meaningful common language can guarantee that players will coordinate on an efficient equilibrium. In figure 4-2 and 4-3, I discussed the possibility for increased coordination in games without communication, by assuming that Pareto-efficient equilibria are natural focal points. I showed that, unless there is a single Pareto-efficient equilibrium that players find focal, this focalness does not itself guarantee that the players will in fact play one of the Pareto-efficient Nash equilibria. To consider the implications of this behavioral assumption generally, I now define a solution concept that assumes that players focus on Pareto-efficient equilibria:

Definition 11. For $i = 1, 2$, let A_i be the set of predictions for player i consisting of the strategy-belief pairs consistent with the set of Pareto-efficient Nash equilibria in the game. *Then Pareto-Focal Rationalizability* is the set of predictions corresponding to the maximal expansion of (A_1, A_2).

We know that Pareto-focal rationalizability is a CBT because of the theorem in the previous section. In figure 4-2, Pareto-focal rationalizability refines rationalizability. In figure 4-3, Pareto-focal rationalizability is equivalent to rationalizability.

We can obviously define the BET based on this CBT:

Definition 12. A Nash equilibrium is a *Pareto-Focal Nash Equilibrium* if it is an equilibrium prediction in the set of Pareto-focal rationalizable predictions.

Because Pareto-focal rationalizability does not refine rationalizability in figure 4-3, Pareto-focal Nash equilibrium is obviously the same as Nash equilibrium in this game. Essentially, we cannot with a behavioral assumption rule out any inefficient Nash equilibria in this game unless we also rule out one of the efficient Nash equilibria. By contrast, while Pareto-focal Nash equilibrium does not guarantee us full efficiency in figure 4-2, we do rule out the most inefficient Nash equilibria. In fact, figure 4-2 well illustrates the fact that the consistency criteria incorporated into BETs in general allow us to refine our predictions in many games, but also tend to restrict us from choosing arbitrary sets of Nash equilibria as theories. The games for which Pareto-focal Nash equilibrium clearly has the most power are those in which there are multiple Nash equilibria, but in which there is a unique Pareto-efficient Nash equilibrium.

Applications of CBTs to communication issues help us see how assuming that players share a common language can lead us to refine Nash equilibrium so that we can predict greater efficiency. Consider my model in Rabin (1991a), which generalizes an example presented in Farrell (1987). This paper posits that two players communicate extensively before they play any given complete-information game. It then posits an assumption about how players use language to focus their behavior. I show that *any* Nash equilibrium in which players use language in this way will be inconsistent with certain inefficient outcomes. But I also show we cannot guarantee that only fully efficient equilibria are played: players might agree to, and play, a Pareto-inefficient Nash equilibrium.

Indeed, from my papers on pre-game communication, and from examining the CBT Pareto-focal rationalizability, a theme emerges. Basically, the general disposition of players to play efficient equilibria does not translate into the ability to play those efficient equilibria, even when they can communicate extensively. Rather, unless there is a uniquely focal efficient equilibrium, we can only guarantee that some of the more inefficient equilibria can be ruled out.

7. Conclusion

This paper has attempted to outline a set of consistency criteria for incorporating behavioral assumptions into formal game-theoretic analysis. Implicit throughout has been the idea that rationalizability and Nash equilibrium are the appropriate nonequilibrium and equilibrium solution concepts that incorporate basic internal-consistency arguments. Yet such a view ignores the many compelling arguments for stronger internal-consistency criteria. These arguments have yielded solution concepts such as iterated weak dominance, and equilibrium concepts such as trembling-hand perfection, sequential equilibrium, and strategic stability.

My approach allows us to incorporate iterated weak dominance into a solution concept; we can begin by eliminating from a game all strategies that do not survive the iterated deletion of weakly dominated strategies. We can then form a hypothetical game by further eliminating strategies based on behavioral assumptions, and apply rationalizability to create a behavioral theory. Then this behavioral theory would pass the test for consistency proposed earlier—that all proposed plans be rational given the overall game—if and only if it would pass the same test where we ignored

	Player 2		
	L	C	R
U	3, 1	1, 0	0, 1
M	0, 0	1, 3	1, 0
D	3, 0	1, 0	1, 2

Figure 4–4.

those strategies eliminated in process of iterated deletion of weakly dominated strategies.

How might we combine equilibrium refinements with behavioral assumptions? One possibility might be to simply combine such refinements in the same way as we combined Nash equilibrium with behavioral assumptions: first we can refine rationalizability using our behavioral assumptions, and then we can select those equilibria in the theory that meet the criteria of the proposed refinement. We could, for instance, simply select the set of perfect equilibria among Pareto-focal rationalizable predictions, rather than the set of Nash equilibria.

Though this approach may be promising, it can be problematic. Consider figure 4-4, and suppose that we construct a CBT by applying rationalizability to the hypothetical game $\{U, M\} \times \{L, C\}$. In this game, rationalizability would eliminate nothing. Thus, the corresponding BET would include two Nash equilibria, (U, L) and (M, C).

What would we get if we applied trembling-hand perfection here? We must be careful to specify whether we would allow "trembles" to be over all strategies in the game, or simply over the strategies in the hypothetical game. If we allowed trembles over all strategies in the game, *neither* of these Nash equilibria are trembling-hand perfect. In general, because our initial internal-consistency test for constructing the CBT did not incorporate the same test we are applying to eliminate equilibria, there is no reason that existence should be guaranteed.

Of course, we could apply the trembles only to those strategies making up the hypothetical game. In figure 4-4, this would mean that we would select only (U, L). Yet this is awkward—seemingly we have made a refinement among the behaviorally plausible strategies based on an internal-consistency argument that we have ignored when testing the overall consistency of the behavioral assumption itself.

The proper approach to combining internal-consistency based refinements with behavioral assumptions would be to somehow incorporate the same criteria motivating the equilibrium refinement into the test for the consistency of CBTs. Essentially, we can define sets ($\mathcal{P}_1^{**}$, $\mathcal{P}_2^{**}$) that—instead of having the simple best-response property incorporated into the set of rational predictions ($\mathcal{P}_1^*$, $\mathcal{P}_2^*$)—incorporate a nonequilibrium version of the consistency criteria incorporated into the proposed equilibrium refinement. Our consistency criterion for a proposed behavioral theory could then be simply whether it is a subset of ($\mathcal{P}_1^{**}$, $\mathcal{P}_2^{**}$).

Notes

1. By prevalent, I include solution concepts such as Nash equilibrium, Selten (1975), Kreps and Wilson (1982), Cho and Kreps (1987), and Kohlberg and Mertens (1986), and Bernheim (1984), and Pearce (1984). Harsanyi and Selten (1988) argue for a theory of rational play that predicts a unique outcome in every game. In my view, theirs is a combination of a theory of rationality and a behavioral theory. Minimally, a solution concept purporting to rely on rationality alone can never rule out strict Nash equilibria. *If* players believed with common knowledge that a particular strict Nash equilibrium would be played, no theory of individual rationality could dictate that any player would wish to deviate. Harsanyi and Selten's algorithm for equilibrium selection rules out strict Nash equilibria.

2. This literature includes Crawford and Sobel (1982), Farrell (1987, 1988, 1991), Myerson (1986), Rabin (1990, 1991), and many others.

3. Could such a conclusion be squeezed out of some internal-consistency arguments based on the payoffs and the structure of the game? I would argue No, and I believe there is a simple "proof" of this perspective. Importantly, nothing in the structure of the game or in the payoffs says that the players speak a common language. Suppose they do not. Would our prediction be that the players would coordinate? No—we invoke the behavioral assumption of meaningful communication only if we believe that the players share a common language. But whether or not the players share a common language appears nowhere in the traditional description of the game; we *must* invoke it as an additional assumption, and its validity cannot be inferred from the payoffs.

4. Throughout the paper, "Pareto-efficient equilibria" means Pareto-efficient among the set of Nash equilibria.

5. The consistency criteria I define correspond to those I have defined earlier in Rabin (1989), and are similar to those defined in Basu and Weibull (1990) and Gul (1991). Moreover, they are the criteria used in Rabin (1990, 1991a, 1991b) in defining communicational solution concepts. Papers that incorporate behavioral restrictions along with rationalizability in noncommunicational settings include Cho (1992) and Watson (1992a, 1992b).

6. The formal development in this paper is for two-person, finite-action games. It could readily be extended to a broader setting, with some complications. Also, while I shall not concentrate on strategies as type-contingent, the approach outlined here can readily be used for incomplete-information games.

7. In the next section, I will present theories which, unlike rationalizability, may exclude some optimal responses by players. In such cases, the force of this condition is that we exclude certain strategies for some beliefs only if we exclude them for all beliefs.

8. The proofs of most of the propositions in this paper follow very closely from the

definitions, and from known features of rationalizability and Nash equilibrium. When this is the case, formal proofs are omitted.

9. Together, these properties constitute the definition of a *regular public theory*, as defined in Rabin (1989).

10. Selten (1975), Kreps and Wilson (1982), and Kohlberg and Mertens (1986) are examples of internal-consistency refinements. Fudenberg and Kreps (1988) is an example of a learning-based refinement of Nash equilibrium. My view is that signaling refinements such as Cho and Kreps (1987) and Banks and Sobel (1987) are implicitly dynamically-motivated refinements.

11. This follows immediately from the convexity condition incorporated into CBTs, because the mixed strategy involves each player believing that the other is playing a convex combination of the other player's pure-strategy-Nash-equilibrium strategies.

12. Rabin (1989) provides an algorithm for checking whether a set of equilibria corresponds to all the Nash equilibria from an CBT. The method involves a generalization of the algorithm developed in the next section.

13. It is fairly straightforward to see that this process will end in a finite number of iterations in any finite-action game.

14. This theme underlies most attempts to incorporate behavioral assumptions. My papers on communication [Rabin (1990, 1991a, 1991b)], and papers such as Cho (1992) and Watson (1991) also seem to relate to the theme of efficiency.

References

Aumann, Robert. 1976. "Agreeing to Disagree." *Annals of Statistics* 4: 1236–1239.

Banks, Jeffrey S., and Randall L. Calvert. 1989. "Communication and Efficiency in Coordination Games." Department of Political Science, University of Rochester, Working Paper No. 196, August.

Banks, Jeffrey S., and Joel Sobel. 1987. "Equilibrium Selection in Signaling Games." *Econometrica* 55: 647–662.

Basu, Kaushik, and Jorgen W. Weibull. 1990. "Strategy Subsets Closed Under Rational Behavior." Olin Program Discussion Paper #62. Woodrow Wilson School, Princeton University, October.

Bernheim, B. Douglas. 1984. "Rationalizable Strategic Behavior." *Econometrica* 52 (July): 1007–1028.

Cho, In-Koo. 1992. "Stationarity, Rationalizability, and Bargaining." The University of Chicago Economics Working Paper Number 92–127, May.

Cho, In-Koo, and David M. Kreps. 1987. "Signaling Games and Stable Equilibria." *Quarterly Journal of Economics* 102 (May): 179–221.

Crawford, Vincent P. 1985. "Learning Behavior and Mixed-Strategy Nash Equilibria." *Journal of Economic Behavior and Organization* 6.

Crawford, Vincent P., and Hans Haller. 1987. "Learning How to Cooperate: Optimal Play in Repeated Coordination Games." mimeo, August.

Crawford, Vincent P., and Joel Sobel. 1982. "Strategic Information Transmission." *Econometrica* 50 (November): 1431–1451.

Farrell, Joseph. 1987. "Cheap Talk, Coordination, and Entry." *RAND Journal of Economics* 18 (Spring): 34–39.

Farrell, Joseph. 1988. "Communication, Coordination, and Nash Equilibrium." *Economic Letters* 27: 209–214.

Farrell, Joseph. Forthcoming. "Meaning and Credibility in Cheap Talk Games." In M. Demster (ed.), *Mathematical Models in Economics*. Oxford: Oxford University Press.

Fudenberg, Drew, and David Kreps. 1988. *A Theory of Learning, Experimentation, and Equilibrium in Games*. Manuscript, MIT and Stanford.

Gul, Faruk. 1991. "Rationality and Coherent Theories of Strategic Behavior." manuscript, Stanford University.

Harsanyi, John C., and Reinhard Selten. 1988. *A General Theory of Equilibrium Selection in Games*. Cambridge, MA: The MIT Press.

Kohlberg, E., and J.F. Mertens. "On the Strategic Stability of Equilibria." *Econometrica* 54: 1003–1038.

Kreps, David, and Robert Wilson. 1988. "Sequential Equilibrium." *Econometrica* 50: 863–894.

Milgrom, Paul, and John Roberts. 1989. "Rationalizability, Learning and Equilibrium in Games with Strategic Complementarities." mimeo, Stanford University.

Myerson, Roger B. 1978. "Refinements of the Nash Equilibrium Concept." *International Journal of Game Theory* 7: 73–80.

Myerson, Roger B. 1986. "Credible Negotiation Statements and Coherent Plans." Discussion Paper No. 691, Center for Mathematical Studies in Economics and Management Science, Northwestern University, August.

Myerson, Roger. 1986. "Multistage Games With Communication." *Econometrica* 54: 323–358.

Pearce, David. 1984. "Rationalizable Strategic Behavior and the Problem of Perfection." *Econometrica* 52 (July): 1029–1050.

Rabin, Matthew. 1989. "Predictions and Solution Concepts In Non-Cooperative Games." PhD Dissertation, MIT, June.

Rabin, Matthew. 1990. "Communication Between Rational Agents." *Journal of Economic Theory*.

Rabin, Matthew. 1991. "A Model of Pre-Game Communication." *Journal of Economic Theory*, forthcoming.

Rabin, Matthew. 1991. "Focal Points and Pre-Game Communication." University of California, Berkeley, Department of Economics Working Paper No. 91-179, September.

Rabin, Matthew. Forthcoming. Corriegum for "Communication between Rational Agents." *JET*.

Selten Reinhard. 1975. "Re-examination of the Perfectness Concept for Equilibrium Points in Extensive Games." *International Journal of Game Theory* 4: 25–55.

Watson, Joel. 1992. "A 'Reputation' Refinement without Equilibrium." mimeo, Stanford University.

Watson, Joel. 1992. "Reputation and Outcome Selection in Perturbed Supergames: An Intuitive, Behavioral Approach." mimeo, Stanford University.

5 ON THE CONCEPTS OF STRATEGY AND EQUILIBRIUM IN DISCOUNTED REPEATED GAMES

William Stanford

1. Introduction

If the theory of discounted repeated games is to accurately represent any real phenomena, then it matters how we model the fundamental concept of strategy in our analysis. This is clearly so because the set of objects we choose to call strategies will in part determine the set of feasible equilibria (jointly rational vectors of strategies) that are consistent with the theory. In turn, these equilibria give rise to sequences of single-period game action vectors that form the outcome paths we expect to observe through time. Thus if we observe an outcome path that is inconsistent with the theory, we must accept either that asserted behavioral postulate of equilibrium is refuted or that there is some other error in our description of the system. Simple examples of the latter include the possibility that some single-period game payoffs have been incorrectly specified or that we have assigned the wrong discount parameter to some player. As indicated above,

---

The author would like to thank James Friedman for many helpful comments.

I will attempt in this chapter to address different modeling issues: how the definition of *strategy* affects our notion of equilibrium and thus how it can affect the set of feasible equilibria.

There are at least two competing paths taken by various researchers in addressing the question of strategy definition. These paths are certainly related since what we might call the first path leads to a definition that formally comprehends the definition of the second. Given a repeated game then, we might deal with either of two sets of strategies with one being a proper subset of the other. As it turns out, some researchers find the restrictive definition more intuitive and so more reasonable. However, in this chapter I will argue that insisting on an equilibrium concept appropriate to this type of limited strategy choice leads to a deficient model. It is deficient in the sense of being incapable of predicting reasonable outcome paths that also accord well with experimental data. In particular, the possibility of predicting cooperative behavior in discounted prisoners' dilemma games is ruled out in this case. I conclude that thoughtful and reasonable players must think in terms of the comprehensive definition and must expect that other players will do so as well.

In the remainder of the chapter, section 2 discusses these issues more specifically. Section 3 contains formal results that are central to the thesis I have outlined above. Finally, section 4 indicates how the results of section 3 can be generalized and contains summarizing remarks.

2. Discussion

There appear to be two somewhat natural approaches to the question of strategy definition in repeated games. First, we may require a strategy to specify actions after all histories of play. In particular, this approach requires actions to be specified even after histories that are inconsistent with the prescribed actions of the strategy itself. Further, if we are interested in subgame perfect Nash equilibria, such a strategy will depend in a nontrivial way on the prior actions of the player employing it. In the conventional interpretation, choosing this kind of strategy reflects a player's concern that mistakes might somehow be made in implementing the actions dictated by the strategy. Such a player might be seen as preparing a defense against the consequences of his own errors, depending on the anticipated reactions of opponents. The most important reference concerning this approach is probably Friedman (1971).

Second, as pointed out in Rubinstein (1991), a strategy may represent a *plan of action*. This phrase seems to be consistent with Shubik's (1982) "complete description of how a player intends to play a game, from beginning to end." The emphasis on players' intentions leads to an approach that limits attention to strategies basing current period action for a player only on the prior actions of opponents. Such a strategy ignores the history of that player's own actions in the repeated game. Thus in one interpretation, a player never seriously allows for the possibility of histories that are inconsistent with the player's own strategy. Presumably the player adopting such a strategy believes his intentions will always be realized and that a defense against the consequences of his own errors will not be necessary. As Rubinstein emphasizes, if we are only investigating Nash equilibria (with no subgame perfection requirements) in an extensive game, then a more broad strategy conception is not necessary. We only check for optimizing induced strategy behavior along a proposed equilibrium path; the consideration of rationality issues concerning a player after he deviates is abandoned. In fact, in the standard definition of Nash equilibrium there is no requirement for optimizing behavior by *any* player in the subgame following some player's deviation. We simply ignore these issues as they concern both a deviator and his opponents. With regard to the definition of strategy, the approach discussed above is taken in Friedman (1968), Rubinstein (1986), Abreu and Rubinstein (1988), Aumann (1981), and Kreps, et al. (1982), for example. In the following, a strategy of this type will be called *reactive*.[1]

The desertion of all rationality requirements following a deviation seems extreme, however. It has been observed elsewhere that if players somehow arrive at a common understanding of how the game will be played then this set of strategy choices and beliefs must constitute a Nash equilibrium. Otherwise the role some player assumes in the understanding is simply not believable. Further, in repeated games (or generally in extensive games) where players threaten effective retaliation against deviations as part of the understanding, these threats must remain credible after a deviation occurs. Otherwise, it remains true that some player's acceptance of the original agreement must be suspect.

All of this leads to more strict rationality requirements than we find in simple Nash equilibria. When we entertain only the possibility of reactive strategies, an apparently reasonable additional rationality requirement is the *paraperfect equilibrium* of Marschak and Selten (1978). A paraperfect equilibrium is described as follows: beginning with a T-period history of play $h(T)$ and a strategy combination f, we say $h(T)$ is *normal for player i with respect to f* if player i has not deviated from the individual strategy f_i in the sequence $h(T)$. Further, we say that a subgame *normal for i with respect to f* lies ahead. Then f is *paraperfect* if for all i and in every subgame normal for player i, f induces a best reply for i to the strategies that f induces for the other players. Paraperfect equilibria are obviously Nash equilibria of the repeated game. Since reactive strategies may be used when a player does not seriously plan on the possibility of his own deviations, all we can reasonably expect is rationality after histories normal for that player. Expecting optimizing behavior of a player after he deviates from a reactive strategy seems hopeless and we will not seek it in results we consider later in this chapter. Thus tentatively at least, the rationality requirements of paraperfect equilibrium and the limitations of reactive strategies appear to be compatible.

At this point, the discussion can proceed in either of two directions. First, we might investigate the relationship between reactive strategies and the requirements of paraperfect equilibrium in discounted repeated games. If we grant the desirability of retaining the "plan of action" interpretation of extensive game strategies then this investigation is one avenue open to us. Alternatively, we will first follow Rubinstein (1991) a bit further in his analysis and then return to the first question in the next Section. To accomplish this consider a case where player one (P1) has just deviated from his strategy for the first time. Checking the optimality of player two's (P2) induced response requires P2 to specify an induced strategy for P1. Consistency now requires P2's beliefs about P1 to constitute P1's actual intended course of future play. To quote Rubinstein, "a strategy encompasses not only the player's plan, but also his opponents' beliefs in the event that he does not follow that plan." This gives rise to the objection that a player's "choice of strategy" should not and does not depend on opponents' beliefs. Rubinstein concludes that "Game-theoretic models which use the notion of subgame perfect equilibrium (or any other solution concept which includes a sequential rationality requirement) therefore require reexamination." In essence, Rubinstein seems to be challenging as unreasonable the requirement that a player should be able to correctly assess opponents' intentions especially after one or more of them have taken an unexpected action.

But I am skeptical of that position and I would try to support that skepticism on the basis that choice of a strategy and opponents' beliefs about that strategy are often not independent phenomena. In broad terms, preplay communications have a key role here simply because they give a context in which intentions and beliefs might be formed and reconciled. More specifically, in serious preplay discussions, the players are trying to reason their way to an understanding of how the game will be played.[2] If this process is to move toward an understanding in any sense, players must develop beliefs concerning opponents' reactions to actions whether or not these actions will later be viewed as deviations from the complete and final agreement. Any confidence in these beliefs must be based on a common knowledge of players' possibilities for action and payoff prospects after any named action. Thus it appears that players must consider the consequences of all their prospective actions in all contingencies for all players along any path to full agreement. In fact, the breakdown in the dichotomy between a player's choice of strategy and opponents' beliefs is the end product of this agreement.[3] It may not be an exaggeration to say that the human species must have developed spoken language and other communication skills in large part to address just such issues. Coordinating intended actions with others' beliefs about those actions is a problem each of us faces and attempts to solve on a daily basis.

As an example, consider some process of communication preceding the play of a discounted prisoners' dilemma. Suppose that in this process, Friedman's (1971) "grim trigger" strategy somehow becomes associated with one of the players, perhaps as the result of focal point effects or because of ingrained personality traits. Since the grim trigger is a dynamic programming best response to itself (with sufficiently little discounting) we immediately have the potential for a full agreement coinciding with a subgame perfect equilibrium of the discounted game. Here, contingencies correspond with a two state system having one "reward" state and one "punishment" state. The consequence of cooperative stage game action by both players is to preserve the present state of the system, while noncooperation by either ensures the system will move to the punishment state.

Finally, the suggestion for reexamination of subgame-perfect equilibrium carries some connotation that the idea and its requirements are flawed or improper, presumably with regard to a large class of extensive games. Merely noting that opponents' beliefs and a player's intentions after he deviates can be reconciled in a natural way might not be enough to resist this conclusion. It may be true for other independent reasons. One would feel much more assured in this resistance in the presence of a convincing equilibrating process for discounted repeated games. Presumably such a process would be based on preplay discussions and lead often and under reasonable conditions to subgame-perfect equilibrium. I am aware of very little research in this area.[4] To summarize, there may be difficult technical problems associated with arriving at some specific subgame perfect equilibrium, but the coordination problem between own intended actions after a deviation and opponents' beliefs appears solvable.

		Player Two	
		N2	C2
Player	N1	0, 0	a_1, b_2
One	C1	a_2, b_1	1, 1

Figure 5–1.

3. Plans of Action and Paraperfect Equilibrium

We return now to the discussion of strategies as plans of action. For this we will consider reactive strategies in the discounted prisoners' dilemma and impose only the rationality requirements of paraperfect equilibrium. This means a player must find it optimal to fulfill threats designed to deter deviations, but no rationality requirements are placed on a player who has temporarily abandoned a reactive strategy. Since a theorem will result from this discussion, we need some formal definitions.[5]

Definition 1. *The basic game we consider has payoffs represented in the two-player matrix shown in figure 5–1.*

We will require $a_1 > 1$, $b_1 > 1$, $a_2 < 0$, $b_2 < 0$, so the game is just the usual prisoners' dilemma. Ci is associated with "cooperation" and Ni with "noncooperation." $G = (S_1, S_2, \pi_1, \pi_2)$ denotes the one-shot simultaneous move game, where $S_i = \{Ci, Ni\}$ is the pure strategy set of player i and $\pi_i : S_1 \times S_2 \to R$ is the payoff to player i as given in the above matrix. Elements of S_i are generally denoted by s_i and are referred to as actions. Also, $s \equiv (s_1, s_2)$ and $S \equiv S_1 \times S_2$.

Definition 2. *Denote by $G^\infty(\alpha, \beta)$ the game with discounting obtained by repeating G countably many times. The discount parameter of P1 is α and β is the discount parameter of P2. A pure strategy for player i is denoted f_i. Such a strategy is a set of functions $f_i = \{f_i(t)\}_{t=1}^\infty$, where $f_i(1) \in S_i$ and for $t \geq 2, f_i(t) : S^{t-1} \to S_i$. The set of repeated game strategies of i will be denoted F_i, and $F \equiv F_1 \times F_2$. Given $f = (f_1, f_2) \in F$, the outcome at time t will be denoted $s(f)(t)$, and defined inductively:*

$$s(f)(1) = f(1) = (f_1(1), f_2(1))$$

and

$$s(f)(t) = f(t)(s(f)(1), \cdots, s(f)(t-1)).$$

The set $\{s(f)(t)\}_{t=1}^{\infty}$ will be referred to as the outcome path induced by f.

Definition 3. *The pair $f \in F$ is a Nash equilibrium if for all $\hat{f}_1 \in F_1$,*

$$\sum_{t=1}^{\infty} \alpha^{t-1} \pi_1(s(\hat{f}_1, f_2)(t)) \leq \sum_{t=1}^{\infty} \alpha^{t-1} \pi_1(s(f)(t))$$

and for all $\hat{f}_2 \in F_2$,

$$\sum_{t=1}^{\infty} \beta^{t-1} \pi_2(s(f_1, \hat{f}_2)(t)) \leq \sum_{t=1}^{\infty} \beta^{t-1} \pi_2(s(f)(t)).$$

Thus if these two conditions hold, each strategy is a best response to the other.

Our convention is to discount to the beginning of period one and we assume that period t payoffs are received at the beginning of period t.

Definition 4. *Given $f \in F$ and $r(1) \in S, \cdots, r(T) \in S$, the pair of strategies induced by f after the history $h(T) = (r(1), \cdots, r(T)) \in S^T$ is denoted $f|_{h(T)} \in F$ and is defined by*

$$(f|_{h(T)})(t)(s(1), \cdots, s(t-1)) = f(T+t)(r(1), \cdots, r(T), s(1), \cdots, s(t-1)),$$

where $s(\tau) \in S$, $\tau = 1, 2, \cdots, t-1$. Denote by $s_i(\tau)$ the projection of $s(\tau)$ onto its i-th coordinate. The pair $f \in F$ is a paraperfect equilibrium if $f_1|_{h(T)}$ is a best response to $f_2|_{h(T)}$ whenever $h(T) = (r(1), \cdots, r(T))$ is a history with $r_1(\tau) = f_1(\tau)(r(1), \cdots, r(\tau-1))$ for $\tau = 1, 2, \cdots, T$ ($h(T)$ is normal for P1 with respect to f) and a similar statement holds with subscripts interchanged.

In the following, we will consider strategy pairs in which at least one of the players bases current period action only on the prior actions of his opponent, ignoring the history of his own actions in the repeated game. To the extent that such strategies capture the "plan of action" idea, the Theorem will show that insisting on this interpretation of repeated game strategies leads to a model with serious deficiencies.[6] In particular, the class of paraperfect reactive equilibria which survive small changes in the players' discount rates consists of trivial equilibria. They are trivial in the sense that both players choose their single shot noncooperative action in each period, all along the outcome paths of all such equilibria. This predicted lack of cooperation in repeated prisoners' dilemma situations is consistent neither with experimental data nor common sense. I would interpret these facts in the following way: with discounting, reactive strategies (and by extension "plans of action") simply have no credibility away from the equilibrium path unless they are trivial. The result depends on requiring rationality only of those who have never deviated. This seems a minimal condition for reasonable human interaction. Thus, rather then reexamining subgame perfect equilibrium notions, I would be more inclined to insist on the full rationality requirements of subgame perfect equilibria. Along with this we need strategies that take account of *all* players' realized prior actions, including those of the player employing any given strategy. Such equilibria are often fully consistent with experimental results indicating that cooperation is a common occurrence.

Returning to the analysis, consider a strategy pair $f = (f_1, f_2)$ that satisfies the following two assumptions:

1. P1 employs a completely arbitrary repeated game strategy $f_1 \in F_1$. P2 employs a reactive strategy $f_2 \in F_2$. Thus $f_2(1) \in S_2$ and for all $t \geq 2$, $f_2(t)(s(1), \cdots, s(t-1)) = f_2(t)(\bar{s}(1)), \cdots, \bar{s}(t-1))$ if $s_1(\tau) = \bar{s}_1(\tau)$ for $\tau = 1, 2, \cdots, t-1$.
2. There exists a set $I_1 \subset (0, 1)$ with a limit point $\alpha \in (0, 1)$ such that for all $\alpha_1 \in I_1$, $f = (f_1, f_2)$ is a paraperfect equilibrium in $G^\infty(\alpha_1, \beta)$.

The first assumption is self explanatory. The second assumption states that the paraperfect equilibrium status of f is reasonably robust to changes in P1's discount rate. The fact that I_1 has a limit point $\alpha \in (0, 1)$ means that every neighborhood of α contains an $\alpha_1 \in I_1$, where $\alpha_1 \neq \alpha$. In particular, I_1 must be infinite, but could consist of an arbitrarily small subinterval of $(0, 1)$, for example. Should something similar to the second assumption fail to hold, we would have a paraperfect equilibrium that is, in a sense, accidental. Such an equilibrium relies on a coincidental and unacceptably sensitive relationship between the payoff matrix entries and the discount parameters of the players. (See Kalai, Samet, and Stanford (1988) for further discussions.)

The following theorem shows that taken together, these two assumptions are quite strong.

Theorem 1. *If the strategy pair $f = (f_1, f_2) \in F$ satisfies Assumptions One and Two then $s(f)(t) = (N_1, N_2)$ for $t = 1, 2, \ldots$.*

Proof. Consider two histories $h(T) = (r(1), \cdots, r(T))$ and $\bar{h}(T) = (\bar{r}(1), \cdots, \bar{r}(T))$ such that $r(1) = \bar{r}(1), \cdots, r(T-1) = \bar{r}(T-1)$ and $r_1(T) = \bar{r}_1(T)$. Assume further that $s(f)(\tau) = (r_1(\tau), r_2(\tau))$ for $\tau = 1, 2, \cdots, T-1$ and $f_1(T)(r(1), \cdots, r(T-1)) = r_1(T)$. Thus both $h(T-1)) = (r(1), \cdots, r(T-1))$ and $\bar{h}(T-1) = (\bar{r}(1), \cdots, \bar{r}(T-1))$ are normal for both players with respect to f and $h(T)$ is normal for P1 with respect to f.

By the definition of induced strategy and the first assumption, the two histories $h(T)$ and $\bar{h}(T)$ induce the same strategy for P2. By the second assumption and the definition of paraperfect equilibrium applied to all $\alpha_1 \in I_1$:

$$\sum_{t=1}^{\infty} \alpha_1^{t-1} \pi_1(s(f|_{h(T)})(t)) = \sum_{t=1}^{\infty} \alpha_1^{t-1} \pi_1(s(f|_{\bar{h}(T)})(t)).$$

In essence, we have two power series which converge to the same function on the set I_1. A well known result from real analysis then yields:

$$\pi_1(s(f|_{h(T)})(t)) = \pi_1(s(f|_{\bar{h}(T)})(t)) \text{ for } t \geq 1. \tag{1}$$

But π_1 is one-to-one on S so that

$$s(f|_{h(T)})(t) = s(f|_{\bar{h}(T)})(t) \text{ for } t \geq 1.$$

Thus

$$\sum_{t=1}^{\infty} \beta_1^{t-1} \pi_2(s(f|_{h(T)})(t)) = \sum_{t=1}^{\infty} \beta_1^{t-1} \pi_2(s(f|_{\bar{h}(T)})(t)). \tag{2}$$

This says that the discounted payoffs to P2 along the induced outcome paths are independent of P2's action in period T. Thus in no period following a history which is normal for both players can the current action of P2 affect his future discounted payoffs along induced outcome paths. Since P2 has chosen a best response strategy (along the equilibrium outcome path), this means that P2 must choose a stage game best response to the prescribed action of P1 in every period along the equilibrium path. Since N2 is strongly dominant for P2, $f_2(t) = N2$ all along the equilibrium path. This easily implies $f_1(t) = N1$ all along the equilibrium path. QED

4. Remarks

There are two key features to the proof of the theorem. The first consists of showing that P2's future discounted payoffs are independent of his own

		Player Two	
		s_{21}	s_{22}
Player One	s_{11}	1, 1	0, 0
	s_{12}	0, 0	1, 2

Figure 5-2.

current action along the equilibrium path (equation 2 in the proof) and thus that P2 must play myopic best response in each period along the equilibrium path. Beyond the two assumptions, this fact was seen to depend on the structure of the payoff matrix to P1: π_1 is one-to-one on $S = S_1 \times S_2$ in the prisoners' dilemma. (As, of course is π_2-all statements in this discussion have their symmetric counterparts.)

In the definitions below we single out two fairly general classes of stage games for which equation 2 continues to hold given the two assumptions. In the following, we assume that S is non-empty and compact, and $\pi_i : S \to R$ is continuous in the product topology for $i = 1, 2$.

Definition 5. *Let $im(\pi_i) = \{\pi_i(s)| s \in S\}$ for $i = 1, 2$. The stage game G will be called a game with completely dependent payoffs if there is a one-to-one and onto map $g : im(\pi_1) \to im(\pi_2)$, satisfying $g(\pi_1(s)) = \pi_2(s)$ for all $s \in S$.*

The prisoners' dilemma is a game with completely dependent payoffs as is, for example, the game derived from the prisoners' dilemma by replicating a row in the payoff matrix, giving P1 an extra strategy. Under the two assumptions and thus given Eq. (1) in the proof of the Theorem, if G has completely dependent payoffs we conclude directly that equation 2 holds.

Definition 6. *The stage game G will be called a game with individually responsive payoffs if for all $s_2 \in S_2$, π_1 is one-to-one on $S_1 \times \{s_2\}$ and for all $s_1 \in S_1$, π_2 is one-to-one on $\{s_1\} \times S_2$.*

Again, the prisoners' dilemma is a game with individually responsive payoffs, but the *augmented* prisoners' dilemma defined as above falls outside this class. Conversely, for a game with individually responsive but *not* completely dependent payoffs, consider a battle of the sexes game (as shown in figure 5-2). Is this case, given $h(T)$ and $\bar{h}(T)$ as in the proof,

Eq. (1) continues to hold. But we also know (since P2 has a reactive strategy) that $(f_2|_{h(T)})(1) = (f_1|_{\bar{h}(T)})(1)$, and thus the fact that payoffs are individually responsive and Eq. (1) force $(f_1|_{h(T)})(1) = (f_1|_{\bar{h}(T)})(1)$. An induction argument now yields Eq. (2).

The second important feature of the proof is its reliance on the existence of strongly dominant strategies for the players. We can relax this to the assumption that G has Nash equilibria, but at a cost. If we strengthen the first assumption so that *both* players adopt reactive strategies and strengthen the second assumption to a symmetric statement, a symmetric argument allows us to conclude that both must play myopic best response in each period. Thus, some stage game Nash equilibrium must be picked out (generally as a function of history) in each period by the strategy pair f, and so collusive outcome paths are ruled out.

Obviously, in some cases it will be possible to leave the first assumption unchanged and still prove a relevant result. An example of this is given by the augmented prisoners' dilemma in which P1 has two undominated strategies and P2 retains a strongly dominant strategy in the stage game.

Essentially, we have seen that with discounting, in any nontrivial paraperfect equilibrium, the strategy of any player depends in part on the prior action history of that player. This is so simply because other players care about the prior actions of that player. I believe the "plan of action" interpretation of repeated game strategies is deficient for just this reason.

In subgame perfect equilibria, this dependence is compounded and sharpened. The problem of coordinating intentions and beliefs is strongly emphasized in Kalai and Stanford (1988), for example. They note that a subgame perfect equilibrium requires the players to find themselves in an equilibrium state in all possible contingencies. Thus in standard repeated game models, there must be agreement among the players concerning the important aspects of a given history and what should be done by all players in light of that agreement. When we model repeated game strategies as simple computing machines (finite automata), this leads to the result that players must choose "isomorphic" machines in any subgame perfect equilibrium.

Finally, I would note that in repeated games without discounting, the problems associated with interpreting strategies as plans of action may be attenuated. In particular (see Stanford, 1986), even though reactive strategies depend directly only on opponents' prior actions, they literally have time to take account of both opponents' and a player's own role in the outcome. With a one period lag, each player is allowed to gauge the effects of his actions on his opponents' behavior. When players ignore the timing aspects of an outcome, a reactive strategy can adequately perform the coordinating function between own actions and opponents' reactions in "real time", as the game progresses. In contrast with the case of discounting, the result may be a subgame perfect equilibrium in reaction functions.

Notes

1. Another approach in which the strategy of any given player is simply undefined after histories inconsistent with it will yield a set of equilibria isomorphic to the set of nash equilibria in reactive strategies.
2. Binmore (1987) discusses more general issues. He examines an "eductive" equilibrating process "which takes place in *notional* time as a result of careful reasoning on the part of the players." While preplay communication has some role in his analysis, he adheres to the "Nash program" and requires that an attempt be made to model preplay communication as additional moves in a larger game in which players choose their strategies independently.
3. Note that outside some context of shared understanding, it is not even clear what the word *deviation* means. Thus the problem of reconciling a player's intentions and opponents' beliefs must be addressed "along the equilibrium path" so to speak, and not just after a deviation occurs.
4. Stanford (1989, 1990) considers an equilibrating process in the discounted prisoners' dilemma and duopoly games whose end product is subgame-perfect equilibrium, but its interpretation as preplay communication is questionable.
5. The material in this Section generalizes somewhat the results of Kalai, Samet, and Stanford (1988).
6. This is in agreement with Rubinstein (1991) in the sense that he also finds problems with interpreting a strategy as a plan of action, but we arrive at this conclusion by a different path.

References

Abreu, D., and A. Rubinstein. 1988. "The Structure of Nash Equilibrium in Repeated Games with Finite Automata." *Econometrica* 56: 1259–1281.

Aumann, R. 1981: "Survey of Repeated Games." In *Essays in Game Theory and Mathematical Economics in Honor of Oskar Morgenstern*. Mannheim, Wien, Zurich: Bibliographisches Institut.

Binmore, K. 1987. "Modeling Rational Players I." *Economics and Philosophy* 3: 179–214.

Friedman, J. 1968. "Reaction Functions and the Theory of Duopoly." *Review of Economic Studies* 35: 257–272.

Friedman, J. 1971. "A Non-cooperative Equilibrium for Supergames." *Review of Economic Studies* 38: 1–12.

Kalai, E., D. Samet, and W. Stanford. 1988. "A Note on Reactive Equilibria in the Discounted Prisoners's Dilemma and Associated Games. "*International Journal of Game Theory* 17: 177–186.

Kalai, E., and W. Stanford. 1988. "Finite Rationality and Interpersonal Complexity in Repeated Games." *Econometrica* 56: 397–410.

Kreps, D., P. Milgrom, J. Roberts, and R. Wilson. 1982. "Rational Cooperation in the Finitely-Repeated Prisoners' Dilemma." *Journal of Economic Theory* 27: 245–252.

Marschak, T., and R. Selten. 1978. "Restabilizing Responses, Inertia Supergames, and Oligopolistic Equilibria." *Quarterly Journal of Economics* 92: 71–93.

Rubinstein, A. 1986. "Finite Automata Play the Repeated Prisoner's Dilemma. "*Journal of Economic Theory* 39: 83–96.

Rubinstein, A. 1991. "Comments on the Interpretation of Game Theory." *Econometrica* 59: 909–924.

Shubik, M. 1982. *Game Theory in the Social Sciences: Concepts and Solutions.* Cambridge, Mass.: MIT Press.

Stanford, W. 1986. "On Continuous Reaction Function Equilibria in Duopoly Supergames with mean Payoffs." *Journal of Economic Theory* 39: 233–250.

Stanford, W. 1989. "Symmetric Paths and Evolution to Equilibrium in the Discounted Prisoners' Dilemma." *Economics Letters* 31: 139–143.

Stanford, W. 1991. "Prestable Stategies in Discounted Duopoly Games." *Games and Economic Behavior* 3: 129–144.

6 THE "FOLK THEOREM" FOR REPEATED GAMES AND CONTINUOUS DECISION RULES*

James W. Friedman
Larry Samuelson

1. Introduction

In the present paper we discuss three joint papers of ours in which the interplay between game theory and oligopoly has been fruitful for both topics; namely the *folk theorem for repeated games* and related literature in game theory and the *reaction function* literature in oligopoly. Both concern the behavior of economic agents, or players, who interact over time; however, the early developments in both game theory and in oligopoly tended to avoid direct confrontation with the analysis of intertemporal behavior. On the one hand, this can be seen in game theory by noting that games in extensive form, where the temporal structure is in plain view, were traditionally converted to the strategic form where the temporal structure became hidden. The move back in the other direction occurred, probably with some stimulation from the concern in economics with

---

* We are grateful to William Stanford for helpful comments. The usual disclaimer applies.

intertemporal behavior, when it was noticed that there are instances in which a given set of players will play a *game* and then will be seen to repeat that same game at regular intervals over time. Early discussions of what may happen when a game will repeated in this way occur in Luce and Raiffa (1957), Aumann (1960), and Friedman (1971).

On the other hand, formal developments in oligopoly theory were virtually entirely static in character until about the 1960s; however, writers discussed their models in dynamic, intertemporal terms. Cournot (1838), Fellner (1949), and Bowley (1924) are typical examples; Fellner (1949) contains many references. They apparently felt a great concern to explain the temporal interactions of firms, even though their formal models remained resolutely single-shot. Cournot provided the first example when he wrote on the *stability* of his equilibrium in the form of *if firm 1 selects* q_1^1, *then firm 2 will want to select* q_2^1, *but then firm 1 would, seeing this, wish to select* q_1^2 *after which firm 2 would want to select* $q_2^2, \ldots$ Eventually this process will, under suitable assumptions, converge to the Cournot equilibrium, but the intertemporal structure of his model is not spelled out. The oligopoly literature since 1960 developed explicit infinite horizon models in which the behavior of each firm i was characterized by a decision function, called a *reaction function*, giving the firm's period t output, q_{it}, as a continuous function of the vector of observed outputs from period $t-1$, q_{t-1}. In these models a firm's objective function is an infinite stream of discounted profits and it is assumed that, at any time t, each firm knows all past output levels chosen by all firms. Attempts to show the existence of subgame perfect reaction function equilibria in this framework failed, apart from trivial cases. See Friedman (1968, 1976), Robson (1986), and Stanford (1986a, 1986b).

The work on both repeated games and reaction functions seeks, at least in part, to show how cooperative payoff outcomes can be supported by subgame perfect noncooperative equilibrium strategy combinations. Such an aim is partly prompted by a belief that outcomes with higher payoffs than those associated with single shot noncooperative equilibrium are not uncommon in practice. The reaction function literature, however, has been a notable failure at proving existence of equilibrium while the developments in repeated games have led to many equilibrium existence results. A striking difference in the repeated game equilibrium strategies and the strategies in the reaction function literature is that the former utilize discontinuous decision rules while the latter utilizes continuous rules. Repeated game results such as Friedman (1971), Rubinstein (1979), and Fudenberg and Maskin (1986), can be applied directly to oligopoly to show how firms could achieve cooperative outcomes in repeated games; however, the resultant strategies would typically use discontinuous rather than continuous decision rules.

So, on the one hand we have the continuous reaction function equilibria of oligopoly whose existence went long unproved, but which we find intuitively appealing. On the other hand we have the equilibria associated with the folk theorem literature which attain the sort of outcomes that have been associated with the continuous reaction functions, but which employ strategies requiring discontinuous behaviors that we find implausible in many situations. In particular, if there is preplay communication, the discontinuous strategy equilibria are quite plausible; they have the requisite self-enforcing character. Without preplay communication, if players are to achieve the sort of outcomes that are associated with cooperative behavior, it is more plausible to us that they evolve strategies in which drastic actions do not appear. The interpretation of such drastic actions is problematic as are the likely reactions of other players. Gradualism, with small changes in one's behavior inducing gradual alterations in the choices of other players seems more intuitively believable. We lack any theory to explain how players might grope their way to these continuous-decision-rule-equilibria; we encourage the study of them in the spirit of enlarging the theoretical arsenal and with the conjecture that they will find a more secure role in the future.

Recent work of Friedman and Samuelson (1990, 1993, 1993a) has been able to bridge these two strands of work by using the technical approach of the folk theorem literature to prove the existence of continuous reaction function strategies. This has been done in a game theoretic framework so that the results can be used both in oligopoly and in other areas as well. In oligopoly, a great deal of the literature is concerned with price, output, and advertising policy. For these decision variables we argue that continuous decision rules make most sense when firms make their decisions without consulting one another in advance.

In the remainder of the paper, section 2 develops the basic model. Then the reaction function approach is reviewed in section 3, the folk theorem literature is reviewed in section 4, and a synthesis is made in section 5, based on Friedman and Samuelson (1990), that deals with supporting payoff outcomes that Pareto dominate single shot equilibrium payoffs, analogous to *trigger strategy equilibria*. There are some limitations to the results of section 5 that are the topic of section 6 where results on duopoly from Friedman and Samuelson (1993) are reviewed. Section 7 is concerned with the work in Friedman and Samuelson (1993a) which is aimed at generalizing the results of section 5 to show that any attainable individually rational payoff vector of the single shot game can be supported by a subgame perfect equilibrium in which the strategies are based on continuous decision rules. Section 8 contains concluding comments.

2. The Single-Shot Game and Repeated Game Models

A single-shot game is characterized by (N, S, P) where $N = \{1, \cdots, n\}$ is the set of players, S_i is the pure strategy space of player i, $S = \times_{i \in N} S_i$ is the strategy space of the game, and P_i is the (scalar valued) payoff function of player i in each period with $P = (P_1, \cdots, P_n)$. The model is subject to the following assumptions:

Assumption 1. The set of players, $N = \{1, \cdots, n\}$, is finite.

Assumption 2. The strategy space of each player $i \in N$ in each period is $S_i \subset \mathbf{R}^m$ and is compact and convex.

Assumption 3. The single-period payoff function of each player $i \in N$, P_i, is single-valued and continuous on S.

Assumption 4. The single-period payoff function of each player $i \in N$, P_i is quasiconcave on S_i.

These assumptions are commonplace and ensure the existence of noncooperative equilibrium in the model. To use this model as the basis for a repeated game, suppose each player i has a discount parameter $\alpha_i \in (0, 1)$ with $\alpha = (\alpha_1, \cdots, \alpha_n)$. Just as $s \in S$ is a strategy combination in the single shot game, $s_t = (s_{1t}, \cdots, s_{nt}) \in S$ is the *action combination* chosen by the players in time t in the repeated game. Given choices $(s_0, s_1, s_2, \cdots)$ the repeated game payoff of player i is

$$\sum_{t=1}^{\infty} \alpha_i^t P_i(s_t) \qquad (1)$$

Complete information is also assumed throughout this paper and, within each time period, the players choose their actions simultaneously. Let $h_t \equiv (s_0, s_1, \cdots, s_{t-1})$ denote *the history of the game at time t*, and denote the set of possible time t histories by $S^t = \times_{t=0}^{t-1} S$. Complete monitoring is assumed. That is, in each period t the players are assumed to choose actions in full knowledge of the actions chosen by all players in each past period.

A strategy for player i will give the period t action of the player at each time period t as a function of the player's accumulated information. Thus each move can be chosen as a function of the history of the game to that time; a strategy for player i in the repeated game consists of an action from S_i for period 0 plus a sequence of decision rules, one for each time $t > 0$, that associates an element of S_i with each member of the set of histories, S^t. Such a strategy will be denoted $\sigma_i = (s_{i0}, \zeta_{i1}, \zeta_{i2}, \cdots)$ where each ζ_{it} is a function from S^t to S_i. Let Z_{it} denote the set of functions that map S^t into S_i for $t \geq 1$. The strategy space of player i in the repeated game is $\Sigma_i \equiv S_i \times Z_{i1} \times Z_{i2} \times \cdots$. Let $\zeta_t \equiv (\zeta_{1t}, \cdots, \zeta_{nt})$ and let $\Sigma \equiv \times_{i \in N} \Sigma_i$. Any pure strategy combination, $\sigma \in \Sigma$, induces a specific *path of action combinations* (or more briefly, *path*) $\mu(\sigma)$ defined in the obvious way.[1] Thus $\mu(\sigma) = (u_0(\sigma), u_1(\sigma), \cdots)$ where $u_0(\sigma) = s_0$, $u_1(\sigma) = \zeta_1(u_0(\sigma))$, and, generally, $u_t(\sigma) = \zeta_t(u_0(\sigma), u_1(\sigma), \cdots, u_{t-1}(\sigma))$. The repeated game payoff function is given by

$$G_i(\sigma) = \sum_{t=1}^{\infty} \alpha_i^\infty P_i(\mathbf{u}_i(\sigma))$$

and, letting $G = (G_1, \cdots, G_n)$, the game is then specified by $\Gamma = (N, \Sigma, G)$. A strategy combination, σ^*, is a *subgame perfect equilibrium* if the continuation of σ^* is an equilibrium point for the subgame after each history $h_t \in S^t$ (for all t).

3. Oligopoly Reaction Functions

Reaction functions are easily formulated in terms of the strategies of the preceding section. A reaction function is an unchanging decision rule that depends only on the actions of the previous period. Thus, $\zeta_{it} = \zeta_i$ is a function from S to S_i that is the same for all t.[2] The strategy of a player may be written $\sigma_i = (s_{i0}, \zeta_i, \zeta_i, \cdots)$. In choosing S_{it} only s_{t-1} comes into play, and the rule never varies over time. The reaction function literature also assumed that i) the ζ_i were continuous functions, ii) if the firms behaved according to $\zeta = (\zeta_1, \cdots, \zeta_n)$, the path of s_t would converge over time to some s^*, and iii) $P_i(s^*)$ would be larger for each player than the single shot equilibrium payoff. Some of these ideas, particularly (i) and (ii), are latent in Cournot (1838) where they make their first appearance in his discussion of the stability of the Cournot equilibrium in terms of the hypothetical reactions of each firm to the choices of its rival. Much the same is true of the famous review and critique of Cournot by Bertrand (1883). The well known *conjectural variation* discussion of Bowley (1924) also allows for (iii), although none of these writers really get beyond the single shot (i.e., one time period with simultaneous choice) framework. Thus conjectural variations models are close to using reaction functions, but they do not do so, as reaction functions explicitly formulate the actions of one time period as functions of the actions of the preceding time period and they typically endow the firm with the objective of discounted profit maximization.[3]

Fellner (1949) provided a major advance in the theory of oligopolistic behavior by *explicitly* considering behavior over time. He suggests *reaction functions* as strategies for firms. The reaction function of one firm will determine the (e.g.) output level to be selected in period t as a function of the output levels chosen by rivals in period $t-1$. These reaction functions would presumably be continuous, so that a small change in behavior by one's rivals will being about a small change in one's own behavior.

The idea of using reaction functions to specify strategies was pursued in a series of several papers starting with Friedman (1968) and ending with Friedman (1976). Infinite horizon oligopolies with discounting were examined with the aim of showing the existence of nontrivial subgame perfect Nash equilibria in which the strategies of the firms are *continuous* reaction functions.[4] This effort made only partial progress and subsequent work of both Stanford (1986a) and Robson (1986) suggested that the effort was, in fact, doomed. Two contrasts can be drawn with these negative results. First, Stanford (1986b) shows the existence of nontrivial subgame perfect equilibria in continuous reaction functions for infinite horizon models in the absence of discounting. Second, the *discontinuous* reaction functions prescribed by *trigger strategies* can be used to obtain nontrivial subgame perfect equilibria with or without discounting. Trigger strategy equilibria in a general game-theoretic model, with application to oligopoly, first appear in Friedman (1971). These are examined in the next section, as they are part of the folk theorem story.

4. The Folk Theorem for Repeated Games and Related Results

The origin of the *folk theorem for repeated games* is claimed by Aumann (1981) to be unknown, but an early statement appears in Aumann (1960). The folk theorem applies to infinitely repeated games without discounting stating that any feasible and individually rational payoff outcome of the single shot game can be the observed payoff outcome of a noncooperative equilibrium of the repeated game. $P(s)$ is *individually rational* if $P_i(s) > v_i$ for $i \in N$ where $v_i = \min_{s_{-i} \in S \setminus S_i} \max_{s_i \in S_i} P_i(s_i, s_{-i})$ is the *minimax payoff of player i* and $v = (v_1, \cdots, v_n)$ is the minimax payoff vector.

Rubinstein (1979) has proven a theorem which is identical to the folk theorem except that it states any individually rational payoff vector can be supported as the observed outcome of a *subgame perfect* noncooperative equilibrium; Rubinstein's result has been extended by Fudenberg and

Maskin (1986) to games with discounting. The proofs of the Rubinstein and the Fudenberg and Maskin theorems have similar flavors which it is instructive to see. The equilibria encompassed by the folk theorem includes *trigger strategy equilibria* which are of interest for their simplicity. Given suitable discount parameters, any single shot payoff that strongly payoff-dominates a single shot noncooperative equilibrium can be supported by a trigger strategy combination, which is a strategy combination of the following sort:[5]

For each player $i \in N$, define σ_i^* as follows: i) At $t = 0$ choose $s_{i0} = s_i^*$. ii) At $t > 0$ choose $s_{it} = s_i^*$ if $s_\tau = s^*$ for $\tau = 0, \cdots, t - 1$. If, at time t, $s_\tau = s^*$ for $\tau = 0, \cdots, t - 1$ does not hold, then choose $s_{it} = s_i^c$.

Thus the players are coordinating on s^* (at which $P(s^*) \gg P(s^c)$ for some single shot noncooperative equilibrium s^c). If any player i at any time deviates from choosing s_i^* then all players $j \in N$ switch to choosing s^c.

Strategies constructed in the proofs of existence theorems by Rubinstein and by Fudenberg and Maskin are considerably more complicated. Those for Fudenberg and Maskin are sketched, as it will be useful to have their strategies in mind in Section 7. Denote by $s^i \in S$ the single shot action combination that minimaxes player i, let $s^* \in S$, let $w^i \in S$ satisfy $P_j(w^i) = x_j^0 + \varepsilon$ for all $i, j \in N$ such that $i \neq j$ and $P_i(w^i) = x_i^0$ for all $i \in N$, $\varepsilon > 0$, and $v \ll x^0 \leq (P_1(s^*) - \varepsilon, \cdots, P_n(s^*) - \varepsilon)$. For each player i there is a *punishment path*, $\mu^i = (s^i, \cdots, s^i, w^i, \cdots, w^i, \cdots) \in \times_{t=0}^\infty S$, in which the first T action combinations are s^i and all remaining action combinations are w^i. There is also an *initial path*, $\mu^0 = (s^*, s^*, \cdots)$ under which s^* is always chosen. The easiest description of Fudenberg-Maskin strategy combination is in terms of these paths:[6]

Players begin by choosing according to the initial path, μ^0. They continue on μ^0 unless a deviation from μ^0 by some player j occurs. If j deviates, then in the period following the deviation they switch to the path μ^j which they start from its beginning (i.e., they choose s^j for T periods, then choose w^j), and they continue on μ^j unless a deviation from μ^j by some player k occurs. If k deviates, then they immediately switch to the path μ^k which they start from its beginning, and they continue on μ^k unless a deviation from $\mu^k \ldots$

Under the Fudenberg-Maskin strategies, a deviator is punished by being minimaxed, but this punishment may be very costly to the punishers. To ensure that the punishers have the necessary incentives, they will achieve much higher payoffs following the T periods of minimaxing (the *punishment phase*) and, if they fail to carry out their punishing role, they will be minimaxed themselves. Note in comparing trigger strategy combinations with what we may call *folk theorem strategy combinations* that the trigger strategy combinations prescribe precisely the same punishment for deviation, no matter who deviates; while the folk theorem strategy combinations punish differently according to which player deviated. Furthermore, it is essential for the folk theorem strategies that the minimaxing duration, T, be finite so that there is a compensating reward for the punishers following the minimaxing phase. Although the Fudenberg and Maskin strategies are much more complex than the trigger strategies, they can support a much larger set of equilibrium payoff outcomes; in general, the minimax payoff vector, v, lies much below the single shot equilibrium payoffs.

There is a final point to be noted here about both trigger strategy combinations and folk theorem strategy combinations: The action taken by a player at time t depends on more than s_{t-1}; it also depends on aspects of the earlier history of play in the game. This contrasts with the reaction function literature which mostly specifies decision rules that make s_{it} a function of only s_{t-1}.[7] We turn now to Sections 5 to 7 where our results are described. Section 5 deals with continuous analogs to the trigger strategy equilibria following Friedman and Samuelson (1990). Section 6 looks at quantity choice duopolies, following Friedman and Samuelson (1993), for which sharper results are obtainable. Then section 7 describes the extension of the results of section 5 to a continuous analog of the folk theorem equilibria following Friedman and Samuelson (1993a).

5. A Continuous Reaction Function Equilibrium

An observation often made about both the trigger strategy equilibria and the folk theorem equilibria described above is that they involve overskill. If a player deviates from what his strategy prescribes, the punishment meted out to the player is the same whether the deviation is huge or tiny. Under the folk theorem equilibria, different players receive different punishments; however, when a particular player is punished, no account is taken of the nature of that player's deviation. This is like sentencing a lawbreaker to fifty years in prison whether he commits brutal murder or parks overtime in a metered spot. In principle, the punishments could be tailored to the size of the deviation, just as they are under legal systems. There are two ways that come immediately to mind for doing this. The first is to reduce the length of the punishment phase; the second is to reduce the severity of punishment during the punishment phase. In the trigger strategy case, the latter would call for choosing something more profitable than s^c following a deviation. This technique is used by Friedman and Samuelson (1990)

THE "FOLK THEOREM" FOR REPEATED GAMES

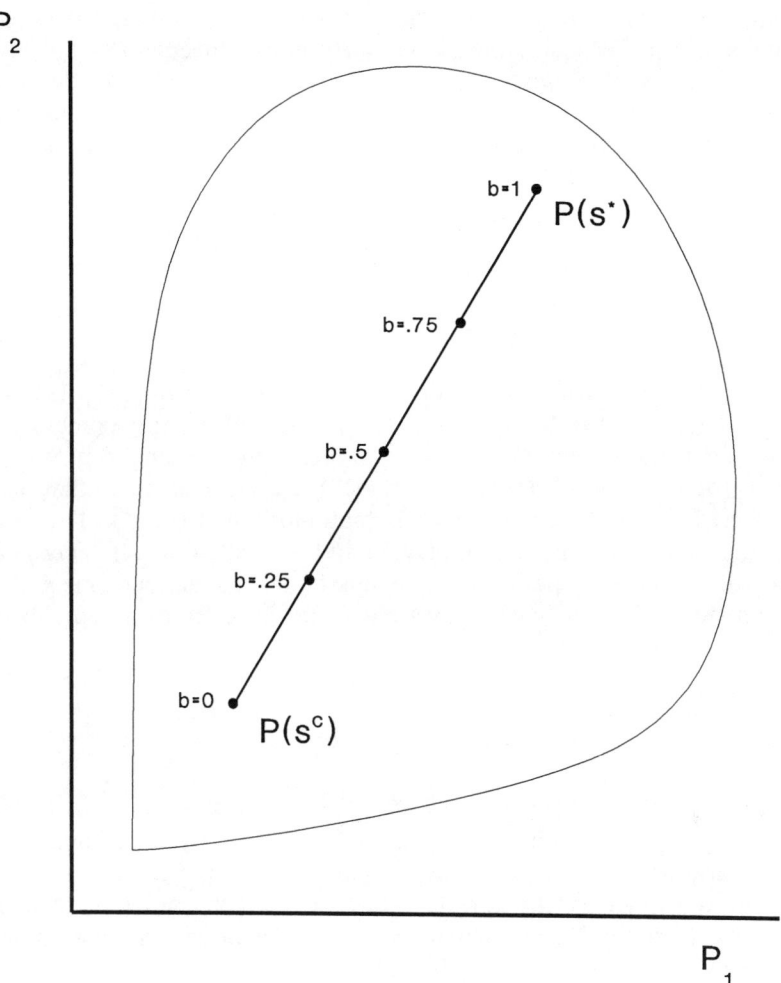

Figure 6–1. A Monotone Path From $P(s^c)$ to (s^*)

and it permits the construction of strategies based on continuous decision rules for the players. Two versions are described below. The key element in both constructions is that, up to a maximal punishment, a player is punished in proportion to the size of her deviation. A tiny deviation receives a tiny punishment; a large deviation receives a large punishment.

Figure 6–1 can be used to describe the latter construction. There is a single shot noncooperative equilibrium payoff, $P(s^c)$, which need not be unique, there is another payoff, $P(s^*)$ at which all players receive higher payoffs than at $P(s^c)$, and there is a straight line connecting $P(s^c)$ to $P(s^*)$. To keep track of points along this straight line, a scale can be drawn assigning the value $b = 0$ to $P(s^c)$, the value $b = 1$ to $P(s^*)$, and giving proportionate values in between. That is, for any s satisfying the condition that

$$\frac{P_i(s) - P_i(s^c)}{P_i(s^*) - P_i(s^c)} = \frac{P_j(s) - P_j(s^c)}{P_j(s^*) - P_j(s^c)} \in [0, 1] \tag{3}$$

$$\frac{P_i(s) - P_i(s^c)}{P_i(s^*) - P_i(s^c)} = \Lambda(s) = b \tag{4}$$

Λ maps the straight line connecting $P(s^*)$ and $P(s^c)$ onto $[0, 1]$. It has an inverse, denoted $\lambda \equiv \Lambda^{-1}$. Thus $\lambda(b)$ is that particular action combination, s, that achieves the point on the straight line corresponding to b. Suppose that $\{\lambda(b) \in S \mid b \in [0, 1]\} = M(s^c, s^*) \subset S$ can be selected so that it is a connected subset of S and so that P is continuous on $M(s^c, s^*)$.[8] Then $\lambda_i(b) \in S_i$ denotes the component of player i and b is called the *reference point*. The value of b in a period corresponds to the payoff outcome for the period that will be achieved if no player deviates from the equilibrium strategies.

The decision rule for player i gives $s_{it} = \zeta_i(s_{t-1}, b_{t-1})$ in the following way: Initially, at time $t = 0$, the reference point is $b_0 = 1$ and $s_{it} = \lambda_i(b_t)$ so that $s_{i0} = \lambda_i(b_0) = s_i^*$. At any time $t > 0$.

$$b_t = \mu b_{t-1} + 1 - \mu \tag{5}$$

if $s_{t-1} = \lambda_i(b_{t-1})$. That is, if $b < 1$ and $s_{t-1} = \lambda_i(b_{t-1})$, the reference point moves slowly upward toward $b = 1$. The rate of movement is governed by μ. It remains to explain exactly how the reference point changes when someone deviates. First the largest actual gain from deviation experienced by a player in period $t - 1$ must be found. It is

$$d_t = \max\{0, P_1(s_{t-1}) - P_1(\lambda(b_{t-1})), \cdots, P_n(s_{t-1}) - P_n(\lambda(b_{t-1}))\} \tag{6}$$

If no deviation occurs or if a deviation occurs that fails to raise any player's payoff, then $d_t = 0$; otherwise, it is equal to the largest gain made by any player. Next let γ be a positive constant which is used to determine a decline in b_t relative to eq. (5) that is proportional to d_t. Thus

$$b_t = \max\{0, \mu(b_{t-1} - \gamma d_t) + 1 - \mu\} \tag{7}$$

So b_t cannot fall below zero, but its value varies continuously with both b_{t-1} and d_t. When there is no deviation, $d_t = 0$ and eq. (7) collapses to eq.

THE "FOLK THEOREM" FOR REPEATED GAMES

(5). Continuity with respect to d_t implies continuity with respect to s_{t-1}. One may summarize eqs. (6) and (7) in the succinct form $b_t = \psi(s_{t-1}, b_{t-1})$. In effect b_t is a behavioral state variable and is an artifact which the players can observe and upon which they condition their behavior, although it does not enter the structure of the payoff functions. In summary

$$s_{it} = \lambda_i(b_t) = \lambda_i(\psi(s_{t-1}, b_{t-1})) = \zeta_i(s_{t-1}, b_{t-1}) \qquad (8)$$

Thus the strategy of player i is $\sigma_i^* = (s_i^*, \zeta_i, \zeta_i, \zeta_i, \cdots)$. All of this can, of course, be repeated for each $i \in N$.

Several questions can be raised about the equilibrium constructed above, such as (1) can this actually be done, (2) what sort of equilibrium does one obtain, (3) roughly speaking, how does the proof work, and (4) how do these decision rules (ζ_i) compare with those in the oligopoly literature? Taking these in order:

(1) There are two problems that might prevent obtaining the desired decision rules. One relates to the requirement that the action combination that achieves a point on the path from $P(s^c)$ to $P(s^*)$ must change continuously as one moves along this path. While it seems intuitively plausible that this be so, and it is true in many examples, nonetheless a general proof is not yet at hand. That is, letting $H(s^c, s^*) = \{P(\lambda(b)) \mid b \in [0, 1]\}$ denote the set of payoffs on the line in figure 6–1 from $P(s^c)$ to $P(s^*)$, there may not be a set of single shot actions, $M(s^c, s^*) \in S$, that maps one-to-one onto $H(s^c, s^*)$ and that is a connected subset of S. Without such a set the ζ_i are not all continuous on S. The second problem relates to whether one can even have an appropriate path from $P(s^c)$ to $P(s^*)$ in the payoff space. It is conceivable that the payoff possibility space has some "holes" in it or has an inconvenient shape as illustrated in figure 6–2. Then some straight line paths would be ruled out. At a minimum, paths are needed along which all payoffs strictly increase as one moves from $P(s^c)$ to $P(s^*)$. In figure 6–2 the shaded regions are not part of the payoff space. As a consequence it is not possible to have a continuous decision rule ($\zeta_i, i \in N$) associated with a subgame perfect equilibrium that attains the payoffs at A or C. Going from $P(s^c)$ to A is impossible on a path whose slope is always positive due to the way the left border of the attainable set bends inward to the right. Going from $P(s^c)$ to C requires a path that either passes above or below the "hole" lying between the points, forcing the connecting curve to have a downward-sloping portion. The payoffs at B are reachable because B and $P(s^c)$ can be connected with a continuous curve along which both payoffs always rise.

(2) When the equilibrium exists, it is subgame perfect. Operationally, this means that whatever the value of b_{t-1} and whatever s_{t-1} is observed,

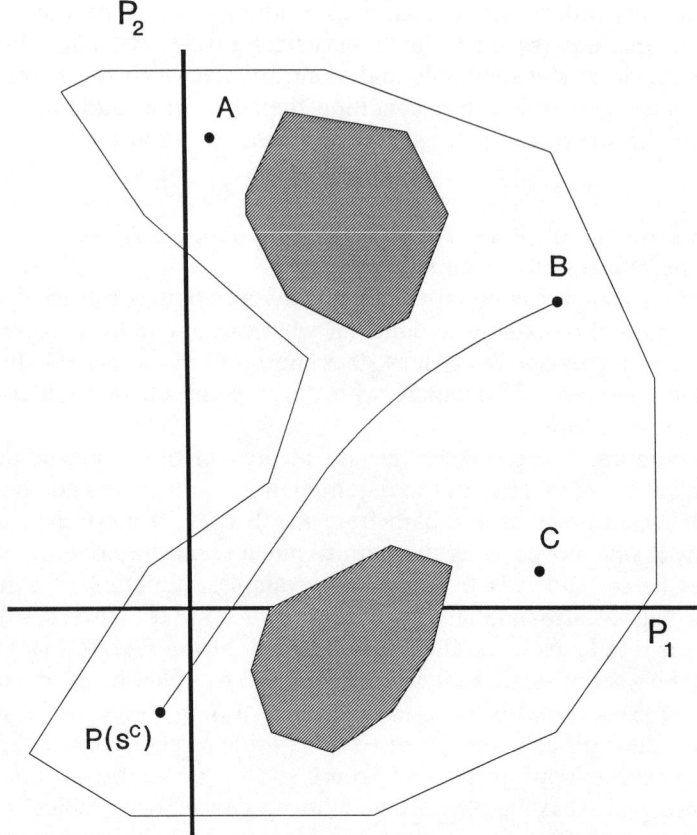

Figure 6–2. A Payoff Space with "Holes"

each player i maximizes his discounted payoff by continuing to choose $s_{it} = \zeta_i(s_{t-1}, b_{t-1})$ as long as the other players ($j \neq i$) are going to choose $s_{jt} = \zeta_j(s_{t-1}, b_{t-1})$.

(3) The proof works in a way that is closely analogous to an existence proof for trigger strategy equilibria. Suppose player i deviates, obtaining an additional d of payoff in the period of deviation. Then the reference point is lowered in all succeeding periods by an amount that decreases over time, so that player i loses on balance if $d < \alpha_i \mu d \gamma/(1 - \alpha_i \mu)$. If $\alpha_i \mu$ is close enough to one, this inequality must be satisfied. Once the reference

point is pushed to $b = 0$, the players are at a single shot noncooperative equilibrium from which deviation is never profitable in the short run. Of course, the key is that, the larger the gain from deviation, the larger is the discounted stream of future payoff reductions for the deviator.

(4) The one outstanding difference between this equilibrium and the reaction function equilibria discussed in the oligopoly literature is that the oligopoly reaction functions do not have a reference point in their formulation. Under oligopoly reaction functions firms choose s_{it} as a function of s_{t-1} alone. Denote such a reaction function for firm i as ϕ_i so that $s_{it}\,\phi_i(s_{t-1})$ and $\phi = (\phi_1 \cdots, \phi_n)$. A reference point provides a form of history dependence that is more complicated than mere dependence of s_{it} on s_{t-1}. Virtually all equilibrium strategies constructed in proofs of the major theorems related to the folk theorem require a form of history dependence that is at least as complex as the history dependence exhibited in the ζ_i above.

The second construction in Friedman and Samuelson (1990) eliminates the reference point, which brings the ζ_i into the same form as the ϕ_i of the reaction function literature. Recall that the purpose of b_{t-1} is to indicate the actions, s_{t-1}, that the players will carry out if no one deviates. That is, the actions called for at time t depend, in general, on the actions *that were supposed to have been carried out* at time $t - 1$ and the actions that were actually carried out at time $t - 1$. Eliminating b_{t-1} from the reaction function requires that the actions the players were supposed to carry out, or some reasonable approximation to them, be inferable from the observed actions. It is possible to eliminate b_{t-1} if an additional condition is used; namely that there is no unilateral Pareto improving deviation possible at any point on the path from $P(s^c)$ to $P(s^*)$. Put another way, let s' be an action corresponding to a point on the path. Then for any $i \in N$ any $s_i \in S_i$, there is at least one player $j \in N$ for whom $P_j(s'\backslash s_i) \leq P_j(s')$. The reason this condition allows the reference point to be dispensed with is that it becomes possible to determine b_{t-1} from s_{t-1} with sufficient accuracy. More exactly, when one or more players deviate, s_{t-1} can be used to determine a value that must be bounded above by the true b_{t-1}. Furthermore, as deviations go to zero, the calculated value converges to the true b_{t-1}.

The noncooperative equilibrium concept is defined in terms of unilateral deviations, not in terms of deviations of two or more players; therefore, a measure of such deviations is needed. In fact it is only necessary to be able to tell when no one deviated and also to tell when a single deviation occurs. Figure 6–3 illustrates how a lower bound can be found for the payoff gain from a unilateral deviation that is at least as large as the actual gain. Suppose the reaction functions (ϕ_i) are to maintain the path from $P(s^c)$ to $P(s^*)$ in figure 6–3 and that the reaction functions direct that

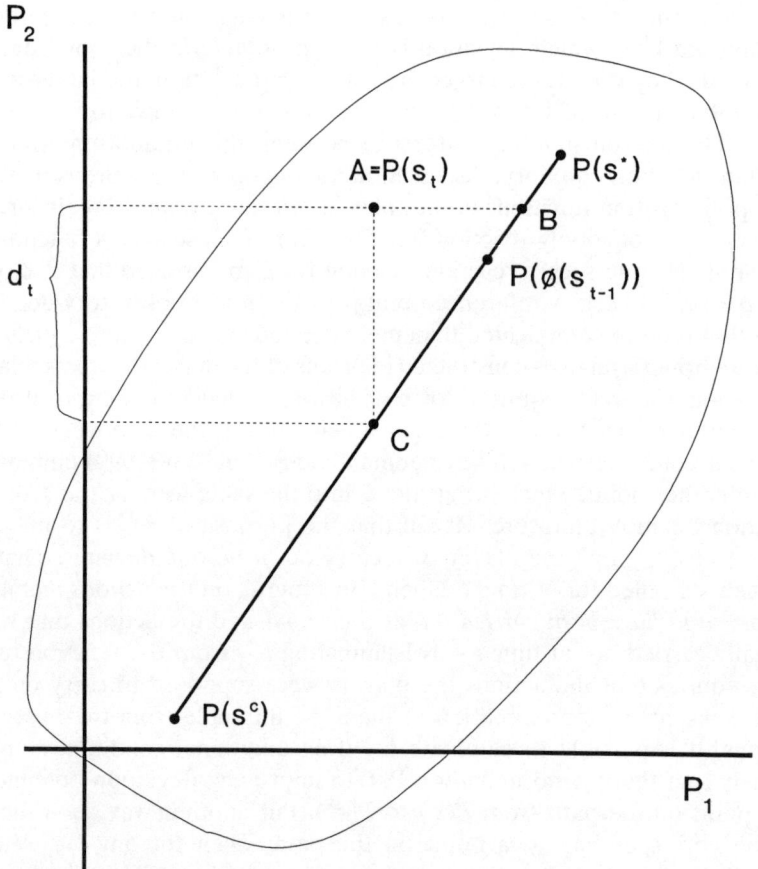

Figure 6–3. Deviations and Implicit Reference Points

$\phi(s_{t-1})$ be selected in period t. Were this done the payoff vector would be $P(\phi(s_{t-1}))$ and the true value of b_{t-1} would correspond to the point $P(\phi(s_{t-1}))$. Now imagine that player 2 deviates, causing the acutal payoff point to be at A. Two points on the path from $P(s^c)$ to $P(s^*)$ are identified by using the payoffs at A. The first is B where the payoff to player 2 is the same as at A. The second is C where the payoff to player 1 is the same as at A. The location of the *implicit* b_{t-1} is the lower of B and C (i.e., at C in this example) while d_t is the difference between the payoff of player 2 at

A and C. These values of b_{t-1} and d_t are used to calculate a provisional b_t, call it b'_t, using equation 7. The action combination $(\lambda(b'_t))$ that achieves b'_t is then defined as $\phi(s_t)$ for the observed s_t that resulted in achieving the payoffs at A. This process can be repeated for any s_{t-1} with the following rules and comment. First, the point A can never lead to a point B or C that is above $P(s^*)$ or below $P(s^c)$. If, for example, $P_2(s_t) > P_2(s^*)$, then B would coincide with $P(s^*)$. Consequently, taking an arbitrary S_{t-1}, it is possible to identify n points on the path from $P(s^c)$ to $P(s^*)$, one for each player. When no deviation has occurred, those n points coincide with $\phi(s_{t-1})$. When a deviation occurs, the largest payoff difference between $P(s_t)$ and the lowest of the n points analogous to B and C provides an upper bound on the extra payoff obtained by the deviator, as well as a lower bound on the location on the path that the players should have attained had no player deviated. This, in turn, provides the value in period t of d_t.

6. A Quantity Choice, Differentiated Products Duopoly

In this section three models of duopoly are examined, taken from Friedman and Samuelson (1993). They are successively more specialized. In the first model, the firms have twice continuously differentiable cost functions with marginal costs that are bounded strictly away from zero. The firms choose output levels with their output pair determining a pair of market clearing prices. Inverse demand functions are nonnegative and twice continuously differentiable. For each firm there is a finite output level so high that its market price will be zero if its output exceeds this bound, irrespective of its rival's output. Inverse demand for a firm is downward sloping in both own and rival's output levels and concave in the firm's own output. The single period profit of a firm is jointly quasiconcave in the two firms' outputs. This latter assumption is not conventional and is somewhat strong; however, the assumptions eliminate the problems noted in section 5. The model *need not be* symmetric.

The assumptions outlined in the preceding paragraph place these models well within the scope of models examined in section 5; however, comparison with the assumptions used in earlier sections will illuminate differences. First, in the present section there are only two players. Second, the model considered in earlier sections did not require differentiability. Third, the duopolists of the present section each have a single decision variable; that is, the single period action set, S_i, is a subset of **R**, the real line, for the duopolists. The assumption of joint quasiconcavity requires special comment. It is made, as we indicate above, to avoid the difficulties that are discussed in section 5 and illustrated by figure 6–2. It is formally a restriction that is not used in the more general treatment found in section 5; however, the theorems in Friedman and Samuelson (1990) are stated to hold under the assumptions given in section 2 *when the difficulties illustrated in figure 6–2 do not impinge*. In other words, we do not know necessary and sufficient conditions that eliminate these problems. For the duopoly model considered below, the joint quasiconcavity assumption is sufficient.

Denote a single shot equilibrium by q^c, the corresponding single shot equilibrium profits by P^c, the set of attainable profits that weakly dominate P^c by H^c, and let P^M be an element of H^c on which the firms coordinate. The assumptions permit proof of the following results. (1) A single shot equilibrium q^c exists at which both firms have strictly positive output levels and strictly positive prices. (2) The points lying above and to the right on any nonnegatively sloped ray through P^c are in H^c until a boundary point of H^c is reached. (3) This boundary point is strongly Pareto optimal. (4) Any Pareto optimal payoff is achieved by a unique output vector and any non-Pareto optimal payoff is achieved by exactly two output vectors. (5) There is a connected subset of output vectors that contains q^c and that map one-to-one onto H^c.

Under the assumptions sketched above, for any $P^M \in H^c$ there exists $\alpha^* \in (0, 1)^2$ such that, if $\alpha \geq \alpha^*$ then P^M can be supported as the outcome in every period of subgame perfect noncooperative equilibrium in stationary, continuous, reaction functions (i.e., reaction functions that map q_{t-1} to q_{it} for each player). These functions can be chosen so that, following any initial choice q_0, the sequence generated by $q_t = \phi(q_{t-1})$ converges to q^M (the output vector that achieves P^M). That is, any single period payoff that can be supported by a subgame perfect trigger strategy combination can also be supported by a subgame perfect reaction function equilibrium. However, suppose some P^M can be supported by trigger strategies when $\alpha = \alpha'$. We do not know whether a reaction function equilibrium can support P^M given α'; what we know is the P^M can be supported for some $\alpha'' \in (0, 1)$.[2] Thus these results for duopoly yield subgame perfect equilibrium in the sort of reaction functions conceived by Fellner (1949) and hinted at by Cournot (1838) and various later writers. Open questions include the possibility of generalizing these results to n firms, using prices rather than output levels as decision variables, and weakening the joint quasiconcavity condition.

Further specializing the model by making the demand functions linear and marginal costs constant, a stronger result can be proved: If some

$P^M \gg P^c$ can be supported by trigger strategies when $\alpha = \alpha'$ then a reaction function equilibrium can support P^M given α'. By making a further specialization of the model to symmetric firms and using Abreu (1986) we can prove that, for suitable discount parameters, any *symmetric* P^M that is individually rational can be supported by a subgame perfect reaction function strategy combination.

7. Supporting Individually Rational Outcomes with Continuous Strategies

The results of Friedman and Samuelson (1990), described in section 5 and amplified for duopoly in section 6, are analogous to trigger strategy formulations (e.g., Friedman, 1971), because they use single shot Nash equilibrium payoffs as the ultimate punishment payoffs and the punishment for deviation does not depend on *who* deviates (although it does depend on the extent of the deviation). They differ from the trigger strategy literature by utilizing continuous decision rules in place of discontinuous rules. In the present section the results of Friedman and Samuelson (1993a) are sketched; these results also utilize continuous decision rules; however, they provide a continuous decision rule version of the Fudenberg and Maskin (1986) extention of the folk theorem. Like Fudenberg and Maskin (1986), a subgame perfect equilibrium supporting an arbitrary individually rational payoff is shown to exist in an infinitely repeated game with discounting.

Recall that strategies in the folk theorem family generally utilize reference points that indicate *where the game is*. For trigger strategy equilibria, there are only two relevant reference point values and they indicate whether there has been prior defection. For the Fudenberg and Maskin theorem, where the game depends upon (1) whether there has been a defection, (2) who is the most recent defector if there has been any defection, and (3) how many periods in the past the most recent defection occurred. These pieces of information are needed to determine the current period action that is called for under a player's equilibrium strategy. In the papers of the *folk theorem literature* (e.g., Friedman (1971), Rubinstein (1979), Fudenberg and Maskin (1986), etc.) the required information from the past is encoded into strategies that determine period t actions as functions of the complete history of play. In Sections 5 and 6, we used a *behavioral state variable* to convey the same information. It takes the form of a reference point with values in the interval [0, 1] that indicates the level of punishment currently being carried out. Zero corresponds to the maximal punishment and one corresponds to no punishment. As with trigger strategies, no account is taken of which player defected.

To carry out a full folk theorem generalization, the scalar reference point must be abandoned in place of a reference point that is a vector of dimension $n + 1$. It is necessary to keep track of the extent to which players are cooperating and also to determine the deviator status of each player. Accordingly, the reference point vector is denoted $b = (b_0, b_1, \cdots, b_n)$. The first coordinate ($b_0$) has the same meaning as our previous reference point and takes values in $[0, 1]$. Thus, b_0 indicates the current *extent of cooperative behavior* with $b_0 = 1$ being maximal cooperation and $b_0 = 0$ being the extreme of punishment. The remaining n coordinates $(b_1, \cdots, b_n)$ designate the degree to which the various players are *current defectors*. $\Sigma_{i=1}^n b_i = 1$ at all times, meaning that there is a nominal defector status across players that always sums to unity. When the game begins, $(b_1, \cdots, b_n) = (1/n, \cdots, 1/n)$ to indicate symmetry of the initial defector status.[9] Thus $(b_1, \cdots, b_n)$ is an element of the unit simplex. Over time if no player ever defects, $(b_1, \cdots, b_n)$ is unchanged. If player i defects, then b_i is increased and the remaining $b_j (j \neq i)$ are decreased in equal proportions to the extent that preserves the total value of unity. A maximal deviation by player i will cause $b_i = 1$ and $b_j = 0, j \neq i$. Similar to Fudenberg and Maskin, if $b_0 = 0$ and, for one player i, $b_i = 1$, then player i is being held to his minimax payoff. As with our earlier formulation, if $b_0 < 1$ and players follow their equilibrium strategies, then b_0 will rise over time, converging to $b_0 = 1$ in the limit. In the following development Assumption 4 from Section 5 is not needed, because the minimax payoffs exist in the absence of the quasiconcavity assumption.[10]

The difficulty here is to make the single period payoff of the players that is prescribed for time t by the decision rules, ζ, a continuous function of b_{t-1} and σ_{t-1} while choosing the decision rules so that any individually rational payoff can be supported. This means moving toward a player's minimax payoff when that player defects as well as moving back toward some initially supported payoff, P^*, following defection. The same problems mentioned in section 5 about "holes," and so forth, arise here and will not be discussed again.

Letting $\Delta_n = \{x \in \mathbf{R}_+^n \mid \Sigma_{i=1}^n x_i = 1\}$ be the unit simplex in $\mathbf{R}^n$, the reference point is $b = (b_0, b_1, \cdots, b_n) \in [0, 1] \times \Delta_n \equiv Y$. The payoffs associated with various reference points can be described by a family of line segments in payoff space, one for each n-vector $(b_1, \cdots, b_n) \in \Delta_n$. These line segments are intimately related to the minimax payoffs of the players and to the payoffs that are sustained when there is no defection. Both s^i and w^i are defined as in section 4. Thus $v_i = P_i(s^i)$ and $P(s^i) = (y_1^i, \cdots, y_{i-1}^i, v_i, y_{i+1}^i, \cdots, y_n^i) \equiv y^i$.

THE "FOLK THEOREM" FOR REPEATED GAMES

Note that we do not know, in general, how a given y_i^j (the payoff player i receives when player j is being minimaxed) is related to v_i. When minimaxing player j, player i could receive a payoff either higher or lower than v_i. Let $v \ll x^0$ and $\varepsilon > 0$. The payoff vector supported by *cooperative* behavior is $\Sigma_{i=1}^n b_i P(w^i)$ under which player i receives $x_i^0 + (1 - b_i)\varepsilon$. Thus if player i has zero defector status ($b_i = 0$) he gets ε more than if he is the sole defector ($b_i = 1$). More generally, given $b \in Y$, the payoff vector called for is $\Sigma_{i=1}^n b_i [b_0 P(w^i) + (1 - b_0) P(s^i)]$. Let $\lambda(b) = (\lambda_1(b), \cdots, \lambda_n(b))$ denote the action combination that achieves this payoff vector. Thus, for $b \in Y$, $\lambda(b) \in S$ satisfies

$$P(\lambda(b)) = \sum_{i=1}^n b_i [b_0 P(w^i) + (1 - b_0) P(s^i)] \tag{9}$$

Let $H(x^0, \varepsilon) = \{x \in \mathbf{R}^n \mid x = \Sigma_{i=1}^n b_i [b_0 P(w^i) + (1 - b_0) P(s^i)], b \in Y\}$. $H(x^0, \varepsilon)$ is the set of payoff vectors that could be chosen by the strategy combinations that are being constructed.

Let $b_t \in Y$ denote the *reference point at time t* and, for $t = 0$, let $b_0 \equiv (1, 1/n, \cdots, 1/n)$. Now the transition of b_t into b_{t+1} can be specified. If no defection took place at time t, then b_{1t} through b_{nt} are unchanged, meaning that the defector status of the players is unchanged and $b_{i,t+1} = b_{it}$ for $i \in N$. If $b_{0t} < 1$, then $b_{0,t+1} = \mu b_{0t} + 1 - \mu$, which is nearer to one. If there was defection, then $b_{i,t+1} > b_{it}$ for the defectors, while, for the nondefectors, $b_{i,t+1} < b_{it}$. Furthermore, the b_i of the nondefectors will be reduced in equal proportions. Also, $b_{0,t+1}$ will be smaller than it would have been in the absence of a defection. To see the preceding in detail, let $P_i^D(s) = \max_{u_i \in S_i} P_i(s \setminus u_i)$ denote the maximum deviation payoff for player i from s. Then z_{it}, defined below, is the normalized extra deviation payoff received by player i if he deviates from $\lambda_i(b_t)$ by choosing some $s_{it} \neq \lambda_i(b_t)$.

$$z_{it} = \max\left\{\frac{P_i(\lambda(b_t) \setminus s_{it}) - P_i(\lambda(b_t))}{P_i^D(\lambda(b_t)) - P_i(\lambda(b_t))}, 0\right\} \tag{10}$$

Note that z_{it} is normalized to measure extra payoff as a fraction of the largest possible extra deviation payoff that the player could have and, if s_{it} reduces the payoff of player i, the extra deviation payoff is taken to be zero. Let $z_t^0 = \Sigma_{i=1}^n z_{it}$, $z_t^M = \max\{1, z_t^0\}$, and $z_t^m = \min\{1, z_t^0\}$. Equations 11 and 12 give the transition from b_t to b_{t+1}.

$$b_{i,t+1} = (1 - z_t^m) b_{it} + z_{it}/z_t^M, \, i \in N \tag{11}$$

$$b_{0,t+1} = \max\{0, \mu b_{0t} - z_t^0 + 1 - \mu\} \tag{10}$$

Equation 11 has the desired property that a player's defector status must be nonnegative, the sum over players is one, and if one or more players defect when the defectors' combined defector status are less than one, their combined defector status rises while that of each nondefector falls to a fraction of its previous value. Equation 12 assures that the level of cooperation is between zero and one and, following defection, it is lower than it would have been in the absence of defection. Note the form of equations 11 and 12 when there is exactly one defector. Then $b_{i,t+1} = (1 - z_{it})b_{it} + z_{it}$ and $b_{0,t+1} = \max\{0, \mu b_{0t} - z_{it} + 1 - \mu\}$. Eqs. (10) to (12), expressing the mapping of (b_t, s_t) to b_{t+1}, can be denoted $\psi: Y \times S \rightarrow Y$. The function ψ is continuous. Using ψ together with λ a decision rule for player i can be specified: $\lambda_i[\psi(b_{t-1}, s_{t-1})] \equiv \zeta_i(b_{t-1}, s_{t-1}) = s_{it}$. This decision rule, along with the period zero action of player i, is the equilibrium strategy of the player. Thus the equilibrium strategy of player i is $\sigma_i^* = (\lambda_i(b_0), \zeta_i)$ and the equilibrium strategy combination is σ^*.

To provide some idea of what can happen in this framework, look at figures 6–4 and 6–5. Figure 6–4 is the situation that is most expected intuitively. The payoff to both players is lower when one is being minimaxed ($b_0 = 0$) than when $b_0 = 1$, irrespective of the values of b_1 and b_2. To see how the game proceeds, suppose that $b_1 = 1$ and $b_2 = 0$. Then, depending on the value of b_0 the players will be on the edge connecting $P(s^1)$ to $P(w^1)$. If $b_0 = 0$ they will be at $P(s^1)$ and if they continue to play with no defections, they will proceed on this edge to $P(w^1)$. Now suppose $b_1 = 2/3$ and $b_2 = 1/3$. They would then be somewhere on the broken line in figure 6–4 running from A to B. If they continue with no defections, they proceed along this broken line toward B. Turning now to figure 6–5, suppose the players are currently at A where $b_1 = 1$. If player 2 should defect, the punishment must be severe enough to avoid permitting player 2 to gain by moving to, say point B, and then proceeding so slowly from B to D that player 2 is better off than he would be in proceeding from A to $P(w^1)$. Thus, it is possible for b_0 to fall and b_2 to rise, as shown in figure 6–5, with the ensuing payoff to player 2 going up in the "punishment" phase as a consequence of his defection. Thus the strategies must be devised so that the payoff to the defector definitely drops. This can be accomplished by making b_2 rise very rapidly as a function of the defection of player 2. For example, a defection by player 2 that reduces b_0 to nearly zero should also raise b_2 enough that the consequent payoff to player 2 falls to beneath his payoff at A.

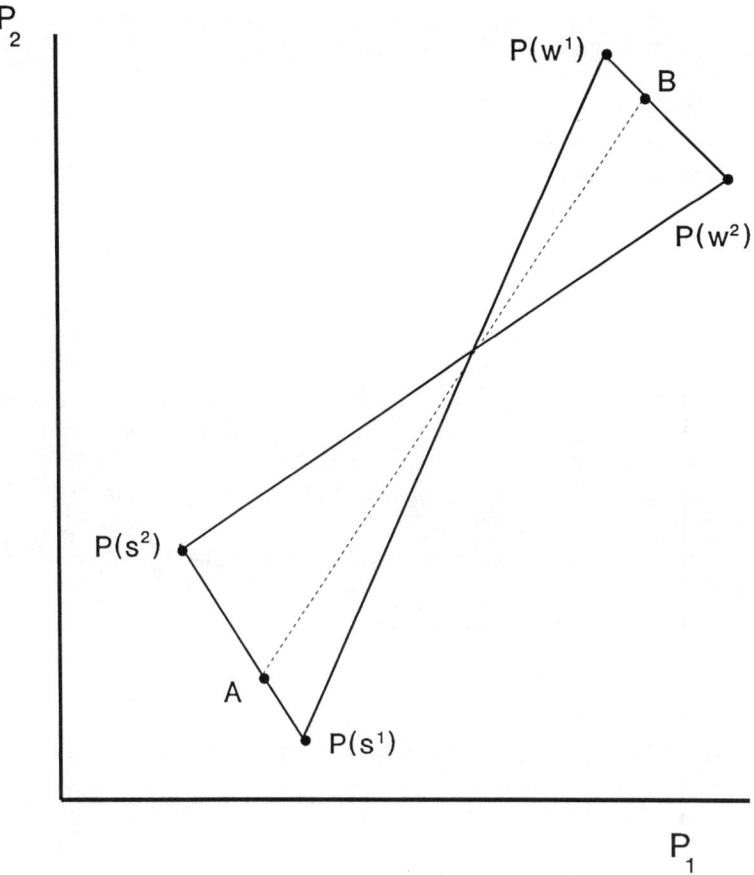

Figure 6–4. Minimax Payoffs Can Be Below Agreement Payoffs

8. Conclusion

We have carried out our work believing that continuous decision rules will sometimes be more plausible than discontinuous and believing that demonstrating the possibility of cooperative outcomes resulting from subgame perfect noncooperative equilibria, as the folk theorem literature demonstrates, does not answer all question of interest. In particular, some strategies that achieve a given outcome may appear more reasonable than

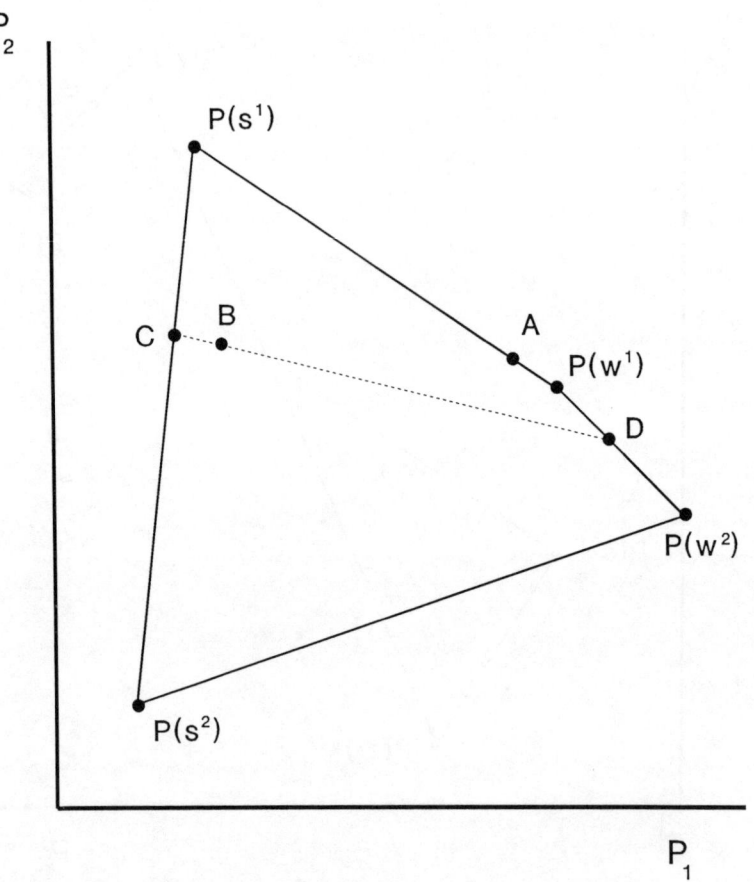

Figure 6–5. Minimax Payoffs Can Be Above Agreement Payoffs

others. These judgments of *reasonableness* are not always subject to verification by proof or by empirical evidence, but they play a role in determining our valuations of research in the discipline.

The results covered in this paper tie together the oligopoly reaction function literature with the folk theorem results in repeated games by providing existence results of the former that use the approach of the latter. Clearly a number of questions remain of which the ability to obtain the set $M(s^c, s^*)$ in the required form is paramount. Two ways around this problem are obvious. One is to require that S maps one-to-one onto the attainable payoff space and to require that each P_i is concave in s. The other is to admit mixed strategies. The first route really indicates obvious sufficient conditions for what we need. Seeing this is no feat; what is needed is to weaken those conditions. We have provisionally adopted the second route.

On balance, we believe the work contained in the papers reviewed here is useful in expanding the nature of potential equilibrium strategy combinations and we hope these formulations will find use, where appropriate, in various applications.

Notes

1. Keep in mind that the sets S_i are sets of pure (nonrandomized) actions and, in general, attention is restricted to pure strategies in this chapter.

2. That is let $h_t = (s'_0, \cdots, s'_t)$ and $h_\tau = (s''_0, \cdots, s''_\tau)$ be any two histories. Then $\zeta_{it}(h_t) = \zeta_{i\tau}(h_\tau)$ whenever $s'_{t-1} = s''_{\tau-1}$. In view of this condition, we may define $\zeta_i = \zeta_{it}$ for all $t \geq 1$ and note that ζ_i maps S into S_i.

3. Indeed, the conjectural variation notion is actually internally inconsistent. It presumes that (a) the players are in a single shot game, each player i selecting s_i exactly once, (b) all players make their choices simultaneously, and (c) each player chooses her s_i after learning what each of the other players have chosen. The difficulties here, of course, are that the players cannot all choose simultaneously, while each one chooses after all the others. Thus the conjectural variation formulations are logically flawed to such an extent that they are nonsense. That the conjectural variation idea came along is understandable, because it is a first attempt to grapple with intertemporal considerations in oligopoly. The remarkable thing is that is has continued to persist in the very recent literature even though intertemporal models have received considerable attention and development.

4. A *trivial* reaction function prescribes precisely the same action at each time t, irrespective of the observed actions of past periods. Trivial subgame perfect reaction function equilibria have long been known to exist: they consist of reaction functions under which each player selects the single shot noncooperative equilibrium action in each period.

5. The trigger strategy combination described below is a *grim trigger strategy combination* or *infinite reversion* strategy combination. We could as well specify finite reversion under which the players choose s^c for a finite number of periods, then return *to* s^*. The grim trigger is easier to discuss because it is simpler.

6. This elegant *path* formulation is due to Abreu (1988).

7. An exception is Marschak and Selten (1978).

8. We address this selection question subsequently.

9. Any other initial values for $b_1, \cdots, b_n$ would be technically feasible; however, common sense suggests that the players begin life symmetrically with respect to deviator status.

10. The v_i must be defined as the supremum over s_{-i} rather than as the maximum and v_i need not be attainable; however, it is possible to get arbitrarily close to v_i.

References

Abreu, Dilip. 1986. "Extremal Equilibria of Oligopolistic Supergames." *Journal of Economic Theory* 39: 191–225.
Abreu, Dilip. 1988. "On the Theory of Infinitely Repeated Games with Discounting." *Econometrica* 56: 383–396.
Aumann, Robert J. 1960. "Acceptable Points in Games of Perfect Information." *Pacific Journal of Mathematics* 10: 381–417.
Aumann, Robert J. 1981. "Survey of Repeated Games." In Aumann et al., *Essays in Game Theory*. Mannheim: Bibliographisches Institut.
Bertrand, Joseph. 1883. Book review of *Théorie Mathématique de la Théorie des Richesses* and of *Recherches sur les Principes Mathématiques de la Théorie des Richesses*. Journal des Savants, 499–508.
Bowley, Arthur L. 1924. *The Mathematical Groundwork of Economics*. New York: Oxford University Press.
Cournot, Augustin. 1838. *Recherches sur les Principes Mathématiques de la Théorie des Richesses*. Translated by N.T. Bacon. New York: Macmillan, 1927.
Fellner, William J. 1949. *Competition Among the Few*. New York: Knopf.
Friedman, James W. 1968. "Reaction Functions and the Theory of Duopoly." *Review of Economic Studies* 35: 257–272.
Friedman, James W. 1971. "A Noncooperative Equilibrium for Supergames." *Review of Economic Studies* 38: 1–12.
Friedman, James W. 1976. "Reaction Functions as Nash Equilibria." *Review of Economic Studies* 43: 83–90.
Friedman, James W. 1990. *Game Theory with Applications to Economics*, 2nd ed. New York: Oxford University Press.
Friedman, James W., and Larry Samuelson. 1990. "Subgame Perfect Equilibrium with Continuous Reaction Functions." *Games and Economic Behavior* 2: 304–324.
Friedman, James W., and Larry Samuelson. 1993. "Continuous Reaction Functions in Duopolies." *Games and Economic Behavior*.
Friedman, James W., and Larry Samuelson. 1993a. "An Extension of the 'Folk Theorem' with Continuous Reaction Functions." *Games and Economic Behavior*.
Fudenberg, Drew, and Eric Maskin. 1986. "The Folk Theorem in Repeated Games with Discounting and with Incomplete Information." *Econometrica* 54: 533–554.
Luce, R. Duncan, and Howard Raiffa. 1957. *Games and Decisions*. New York: Wiley.
Marschak, Thomas, and Reinhard Selten. 1978. "Restabilizing Responses, Inertia Supergames, and Oligopolistic Equilibria." *Quarterly Journal of Economics* 92: 71–93.
Robson, Arthur. 1986. "The Existence of Nash Equilibria in Reaction Functions for Dynamic Models of Oligopoly." *International Economic Review* 27: 539–544.
Rubinstein, Ariel. 1979. "Equilibrium in Supergames with the Overtaking Criterion." *Journal of Economic Theory* 21: 1–9.
Samuelson, Larry. 1987. "Nontrivial Subgame Perfect Duopoly Equilibria Can Be Supported by Continuous Reaction Functions." *Economics Letters* 24: 207–211.

Stanford, William G. 1986a. "Subgame Perfect Reaction Function Equilibria in Discounted Duopoly Supergames are Trivial." *Journal of Economic Theory* 39: 226–232.

Stanford, William G. 1986b. "On Continuous Reaction Function Equilibria in Duopoly Supergames with Mean Payoffs." *Journal of Economic Theory* 39: 233–250.

7 ALTERNATIVE INSTITUTIONS FOR RESOLVING COORDINATION PROBLEMS: EXPERIMENTAL EVIDENCE ON FORWARD INDUCTION AND PREPLAY COMMUNICATION

Russell Cooper, Douglas V. DeJong, Robert Forsythe, and Thomas W. Ross

1. Introduction

As the timely appearance of this volume suggests, the existence of coordination failures in a variety of strategic settings has begun to receive increased attention. Numerous games have been described in which players are required to coordinate their actions in order to reach a mutually advantageous equilibrium. Examples include network externalities (see e.g., Katz and Shapiro, 1985), product warranties with bilateral moral hazard (Cooper and Ross, 1985) and team production (Bryant, 1983). Recent work on macroeconomic models of imperfectly competitive economies (e.g., Heller, 1986, and Cooper and John, 1988) and search (Diamond, 1982) has also identified the possibility of aggregate coordination failures. More

---

This paper has been prepared for inclusion in the forthcoming volume, *Problems of Coordination in Economic Activity*, edited by James W. Friedman. We are grateful to many people who have provided useful advice and criticism throughout the extended period of time covered by this work. Financial support has been provided by the National Science Foundation and the Sciences and Humanities Research Council of Canada.

Coordination Game
Column Player's Strategy

		1	2
Row Player's Strategy	1	800, 800	800, 0
	2	0, 800	1000, 1000

Figure 7–1. Games CG

recently, experimental studies have demonstrated that there are a number of environments in which coordination failures do indeed occur (e.g., Cooper, et al., 1990, hereafter CDFR1, and Van Huyck et al., 1990).

In this paper we distinguish between two types of coordination failures. One is an equilibrium selection problem: when might players unilaterally take actions which would lead to a Pareto-inferior Nash Equilibrium being played? This perspective assumes that players will always achieve an equilibrium outcome. However, a second kind of coordination failure can arise if a pure strategy equilibrium is not attained and ex post disequilibrium outcomes result. It recognizes that in some games, players who disagree on which equilibrium should be chosen might well agree that some disequilibrium outcomes should be avoided.[1]

This paper reviews our earlier experimental work which demonstrates that in certain experimental environments, both kinds of coordination failures will commonly occur.[2] It further examines the extent to which alternative institutions can mitigate these failures. The two alternative institutions are those in which a stage of nonbinding preplay communication is permitted and those in which an outside option is added as a first stage to the game. With communication, we ask whether such "cheap talk" avoids suboptimal play by focusing players' beliefs on a Pareto-dominating equilibrium outcome. With outside options, we examine whether the logic of forward induction holds and thus, coordinates play.

We study two games that illustrate both of these coordination problems; a coordination game and a battle of the sexes game. The two games are illustrated in the following figures. Figure 7–1 displays the coordination game (CG) while figure 7–2 shows the battle of the sexes (BOS).

Game CG has the important property of a coordination game: multiple, Pareto-rankable Nash equilibria, at outcomes associated with strategies (1, 1) and (2, 2). The Pareto dominant equilibrium occurs when both players select strategy 2; any other result is termed a coordination failure. Notice, however, that while players have something to lose by being in the "wrong"

Battle of the Sexes Game
Column Player's Strategy

		1	2
Row Player's Strategy	1	0, 0	200, 600
	2	600, 200	0, 0

Figure 7–2. Game BOS

equilibrium (the first kind of coordination failure), even this equilibrium weakly dominates the two disequilibrium outcomes (the second kind of coordination failure). The possibility of disequilibrium, reflecting strategic uncertainty, could certainly influence the choice of the players. For example, the possibility of reaching the (2, 1) outcome might lead row to play strategy 1 which is safer in the sense that it provides a payoff of 800, independent of column's choice.[3]

Both kinds of coordination failures are also present in game BOS.[4] Here, there are two pure strategy asymmetric Nash equilibria associated with strategy pairs (1, 2) and (2, 1) and one mixed strategy symmetric equilibrium where each player plays strategy 2 with probability 3/4. Examining the pure strategy solutions identifies the elements of conflict; the players rank the two pure strategy equilibria differently, row preferring (2, 1) and column (1, 2). As much as each player would prefer his favorite equilibrium, the "wrong" equilibrium is better for each player than the mixed strategy equilibrium (where expected payoffs are 150 for each player). Similarly, both players prefer either pure strategy equilibrium to either disequilibrium outcome. Due to this, players should be interested in institutions that will help them coordinate their actions.

There is now a great deal of experimental evidence that coordination failures occur.[5] In particular, as discussed below, our work with these two games has revealed frequent failures. In one-shot versions of CG, the Pareto-inferior outcome (1, 1) is by far the most common outcome and in one-shot BOS games play results in ex post disequilibrium outcomes about 60% of the time.

As coordination does not appear to come naturally, it is therefore important to understand how players might try to adapt to the possibility of coordination failures. In particular, are there changes to the basic game which will mitigate coordination problems? The next section describes the effects on the play of these games of the introduction of two possible pre-play institutions: the presence of an outside option for one of the players and of permitting cheap-talk preplay communication. The rejection of an outside option by one player might serve, applying forward induction reasoning, to signal his intended actions to the other player. Similarly, even if preplay communication cannot produce binding agreements, it may help focus beliefs on a particular equilibrium.

Our results indicate that both cheap talk and forward induction can help resolve coordination problems but that the impact of the institution depends upon the game. Coordination failures in CG arise due to the lack of confidence by one player that the other will choose strategy 2. It appears that two-way communication coordinates play on the Pareto-dominant Nash equilibrium. For CG, neither one-way communication nor allowing one player an outside option appears capable of building enough confidence to support the best equilibrium. In contrast, either one-way communication or an outside option appears sufficient to resolve the conflict in BOS. However, two-way communication does not resolve the choice of equilibrium for the players. Apparently, the addition of some asymmetry into this game is sufficient to overcome coordination failures.

2. Overcoming Coordination Problems

The two games we study here have an important property in common: in both, players have an incentive to find their way to one of the pure strategy equilibria rather than to fall into ex post disequilibrium. This suggests a need for coordination in both cases. The games differ, however, in another important respect: the pure strategy equilibria can be Pareto-ranked in the coordination game but not in the battle of the sexes. This fact suggests that the conflict inherent in the second game could lead players to try to use forward induction or preplay communication to their private end.

The institutions studied are also rather similar. Both involve adding a stage prior to the play of the game in which one or both players can make some move that has no direct impact on the strategies available or payoffs in the game to follow. However, in both cases important effects in equilibrium are possible if the "irrelevant" stage changes players' beliefs about how their rivals are going to play.

2.1. Outside Options and Forward Induction

Consider first the potential impact of forward induction reasoning on the play of the BOS game. Suppose the row player was permitted to choose whether to play this game, and should he choose not to play he would collect a payoff of 300. We refer to this as row's outside option and assume that it is common knowledge. What should column infer about row's intentions if she observes him reject the outside option? Since he should only reject the outside option if he expects to earn more playing the game, the logic of forward induction suggests that she can discount completely the prospect that row will play strategy 1 since that cannot lead to a payoff higher than 300. Row must therefore be planning on playing strategy 2 and column should best-respond with her strategy 1 and the payoffs (600, 200) follow. It should be clear that any outside option between 200 and 600 would work as well by this reasoning. Outside options paying less than 200 however should be irrelevant while options paying more than 600 should never be rejected.

The equilibrium selection problem is not solved simply by the fact that we have defined a new game with the addition of the outside option. While it is true that the play of (1, 2) is no longer part of a subgame perfect equilibrium, there are still two such equilibria. The first is that just described: row rejects the option and the (2, 1) outcome is observed in the subgame. The second involves row taking the outside option and is supported by beliefs that column will play 1 with sufficiently high probability. It is this second equilibrium that does not survive forward induction.

In this way, forward induction tells us that the histories players bring to a particular subgame can affect the beliefs their rival's hold about their intended actions. While essentially an extensive form concept, it is easy to see that the logic that drives equilibrium selection here is, in this game at least, equivalent to the iterative deletion of dominated strategies in the normal form.[6]

The argument about the effect of outside options in CG is essentially the same. If the outside option paid row 900 (or any amount between 800 and 1000), column should infer row's intention to play 2 from a rejection of that option. This focuses beliefs on the Pareto-dominant Nash equilibrium and resolves the coordination problem. Outside options paying less than 800 should always be rejected (they are dominated by the play of strategy 1) and are therefore irrelevant, while options paying more than 1000 should never be rejected.

2.2. Preplay Communication

Suppose that, prior to the play of the BOS game, row is permitted to send a message to column. The message in this communication stage is constrained to be a simple announcement of row's intended play, and it does not bind row's choice in the next stage. Assume also that this is all common knowledge. Obviously, row would like to use this cheap talk to help both players avoid ex post disequilibrium and to attain his preferred outcome (i.e. (2, 1)) from the set of equilibria. Whether he can do so will depend upon the meaning players attach to such cheap talk messages.

Farrell (1987) argued that it is reasonable for cheap talk announcements to be taken at face value if it will indeed be optimal for the sender to keep his promise if he expects the receiver to believe the message. In this fashion, one-way cheap talk permits row to select the Nash equilibrium of his choice, namely (2, 1). This way, all coordination problems are resolved. Alternatively, if players believe that cheap talk messages are selected randomly by the sender, no meaning will be attached to them and cheap talk will not influence the play of the game.

If communication is permitted in both directions, the story is slightly more complicated. Again following Farrell (1987), assume that:

(a) If the announcements of both players constitute a pure-strategy Nash equilibrium for the second-stage game, each player will play his announced strategy; and,

(b) If the announcements of both players do not constitute a pure-strategy equilibrium in the second-stage game, each player will behave as if the communication had never happened.

With these assumptions, two-way preplay communication partially resolves coordination problems in BOS. The only symmetric solution to the two stage game is a mixed strategy one, but it does increase payoffs relative to the mixed strategy equilibrium of BOS (see CDFR2).[7]

These assumptions also make communication valuable in CG. For the same reasons, one-way communication should completely resolve the coordination problems, and two-way should help. There are reasons, however, to expect that two-way communication might do even better than one-way in CG. In this game, a question of confidence arises: How confident can row be that column will play strategy 2 given that 1 is "safer"? It may be that two-way communication is a superior way to build the confidence necessary for both players to play 2.

In what follows, we report results on the following ten treatments:

1. CG   the coordination game from figure 7–1
2. CG–900   game CG with an outside option paying (900, 900)

3. CG-700   game CG with an outside option paying (700, 700)
4. CG-1W   game CG with one-way cheap talk communication
5. CG-2W   game CG with two-way cheap talk communication
6. BOS   the battle of the sexes game from figure 7–2
7. BOS–300   game BOS with an outside option paying (300, 300)
8. BOS-100   game BOS with an outside option paying (100, 100)
9. BOS-1W   game BOS with one-way cheap talk communication
10. BOS-2W   game BOS with two-way cheap talk communication

We will examine not only how well the two institutions did in resolving coordination problems in the two games but also the role of focal point effects in generating these outcomes. In the BOS game, both players prefer either of the pure strategy equilibria to the mixed strategy equilibrium. Thus, it is plausible that a one-sided outside option or one-way communication might be able to facilitate coordination in BOS just by their ability to point subjects toward one of the two pure strategy equilibria. It is less clear that this is possible in CG. To test this idea we created a design in which row was given an outside option of such a low level (100 in BOS, 700 in CG) that forward induction arguments cannot be applied to coordinate play. If the focal point effects are present, this "irrelevant" outside option should prove helpful in BOS but not in CG.

3. Experimental Design

The design for these treatments is explained in considerable detail in our previous papers. Here we provide an overview of the procedures. The interested reader should consult the relevant papers for details.

In the experiment, players participated in games such as CG and BOS. Each player was paired with an anonymous opponent. One was designated the row player and the other the column player. Each game was designed to be one of complete information because each player's payoff matrix was common knowledge and the numerical payoffs represented a player's utility if the corresponding strategies were chosen. To accomplish this, we induced payoffs in terms of utility using the Roth-Malouf (1979) procedure.[8] With this procedure, each player's payoff was given in points; these points determined the probability of the player winning a monetary prize. At the end of each period of each game, we conducted a lottery where "winning" players received $1.00 or $2.00, depending on the session and game, and "losing players received $0.00.[9] The probability of winning was given by dividing the points the player had earned by 1000. Since expected utility is invariant with respect to linear transformations, this procedure ensures that when players maximize their expected utility, they maximize the expected number of points in each game regardless of their attitude toward risk.

Each of the ten treatments used three cohorts of players, with each cohort consisting of eleven different players. All players were recruited from upper division undergraduate and graduate classes at the University of Iowa. Upon their arrival, players in a cohort were seated at separate computer terminals and each was given a copy of the instructions for the experiment. These instructions are reproduced in our previous papers. Since these instructions were also read aloud, we assume that the information contained in them is common knowledge.

Each player participated in a sequence of one-shot games against different anonymous opponents within his cohort. All pairing of players was done through the computer using a procedure described below. Since players reported their strategy choices through computer terminals, no player knew the identity of the player with whom he was currently paired, nor the history of decisions made by any of the other players in the cohort.

Each cohort participated in two separate sessions.[10] In Session I, all players participated in ten symmetric one-shot dominant strategy games. The payoff matrix used in each of these games is the same and is given in our previous papers. During Session I, each player played once against every other player. Since there was an odd number of players, one sat out each period. Thus, Session I consisted of eleven periods. Also, players alternated being row and column players during the periods in which they were active participants.[11] Session I was conducted for two reasons: first, to provide players with experience with experimental procedures, and second, to see how well the dominant strategy equilibrium prediction performed.[12]

In Session II, all players participated in twenty additional one-shot games which differed from the game played in Session I. Each played against every other player twice: once as a row player and once as a column player. As in Session I, one player sat out in each period and players alternated between being row and column players during the periods in which they were participating. Thus, Session II consisted of twenty-two periods.[13]

For the baseline treatments, the Session II game was either CG or BOS with the payoffs given in figures 7–1 and 7–2. For the outside option treatments, the payoffs for the outside option were varied across treatments as noted in the previous section. In the game, row choose between the outside option or playing the subgame. Column was informed of row's choice

Table 7–1. Coordination Game: Last 11 periods[1]

Game	outside option	(1, 1)	(2, 2)	(1, 2) or (2, 1)	Total
CG	—	160	0	5	165
		(97%)	(0%)	(3%)	
CG-900	65	2	77	21	165
		(2%)	(77%)	(21%)	
CG-700	20	119	0	26	165
		(82%)	(0%)	(18%)	
CG-1W	—	26	88	51	165
		(16%)	(53%)	(31%)	
CG-2W	—	0	150	15	165
		(0%)	(91%)	(9%)	

[1] Numbers in parentheses refer to proportions of play in the subgame of the outside option treatments.

and in the event the subgame was selected, actions were chosen simultaneously as in the baseline.

For the pre-play communication games, in the one-way communication treatment, row sent a message to column of either 1 or 2. Once the message was received, actions of the players were chosen simultaneously. Two-way communication worked the same way except that in the first stage, both players simultaneously sent messages.

4. Results

To simplify the tabular presentation of the Session II data we report the results by game.[14] As reported in our earlier papers, the results from Session I indicated that play was almost exclusively at the dominant strategy equilibrium, particularly by the end of the session. In evaluating the outcomes from Session II, we present results from the last 11 periods for each treatment.[15]

4.1. CG

Table 7–1 presents the results on the last 11 periods of the coordination games. The first row of the table reveals the extent of the coordination problem: out of 165 outcomes in the last 11 rounds of CG, not one was at

Table 7–2. Coordination Game: Mapping of Announcements to Actions in 1-way Communication Game

Announcement by Row	Actions: Row, Column				
	(1, 1)	(2, 2)	(1, 2)	(2, 1)	Total
1	20	1	0	0	21
2	6	87	22	29	144
Total	26	88	22	29	165

the Pareto-dominant equilibrium (2, 2). Equilibrium play was not an issue, however; 97% of play was at the Pareto-inferior Nash equilibrium (1, 1).

When the game was played with a prior stage in which row had the choice of a "relevant" outside option paying 900, play changed rather dramatically. Conditional on playing the subgame, 77% of outcomes were at the Pareto-dominant equilibrium and only 2% at (1, 1). These results are consistent with forward induction. Contrary to the predictions of forward induction, however, the outside option was selected almost 40% of the time. Recall that this is still part of a subgame perfect Nash equilibrium.

Neither forward induction nor focal point arguments would apply with an irrelevant outside option in this game and this is confirmed by the data in the third row. When the outside option is only 700 (and therefore below the critical level of 800), it is chosen much less often and play conditional on its rejection is not very different from game CG. Most significantly, there are again no (2, 2) outcomes.

The last two rows present the outcomes from the CG communication treatments. One-way communication significantly improves coordination on the dominant equilibrium (now achieved 53% of the time), though the results are still far from those predicted by Farrell's theory. To help illustrate what messages are being sent and how players respond, table 7–2 presents the announcement data from the last 11 periods of this treatment. Over 87% of the time row announces strategy 2 and on over 80% of those occasions he follows through and plays 2. Column seems to understand the message, playing 2 over 75% of the time she hears row announce 2. However this frequency is low enough that playing 1 is still a reasonable strategy for a row player who had announced 2. A message of 1 is even clearer: in 20 out of the 21 times row announced 1, play was at (1, 1). Therefore, the failure to coordinate more frequently on (2, 2) reflects row's failure to announce 2 all the time and the fact that, conditional on an announcement of 2, players do not always play according to the theory.

Table 7-3. Coordination Game: Mapping of Announcements to Actions in 2-way Communication Game

Announcements Row, Column	Actions: Row, Column				
	(1, 1)	(2, 2)	(1, 2)	(2, 1)	Total
1, 1	0	0	0	0	0
2, 2	0	150	7	8	165
1, 2	0	0	0	0	0
2, 1	0	0	0	0	0
Total	0	150	7	8	165

Though one-way communication clearly helped mitigate coordination problems, the single announcement by row was not enough to make players fully confident about the equilibrium to be selected. Notice that the cheap talk message was less successful at coordinating play at (2, 2) than the forward induction "message" sent by rejection of the 900 outside option. However, one-way communication generated 88 plays of (2, 2) while, due to the frequent selection of the outside option, the Pareto-dominant equilibrium is observed only 77 times in CG-900.

It appears that two-way communication does generate an extra level of confidence. As the last row of Table 7-1 reveals, play in CG-2W was remarkably focused on the Pareto-dominant equilibrium: over 90% of outcomes were of (2, 2).

Table 7-3 reports the mapping of announcements to actions in the two-way communication game. Strikingly, every announcement in the last 11 rounds was of strategy 2 and only 9% of the time did anyone then play anything other than strategy 2. Row and column were about equally responsible for these few deviations. Therefore, while the rejection of a relevant outside option might be seen to be a more credible statement of intentions on the part of row than a cheap talk message, the additional confidence provided when both players signal their intentions to play strategies supporting the Pareto-dominant equilibrium makes this the most effective institution for resolving coordination problems in this game.

4.2. BOS

Table 7-4 reports the results on the last 11 periods of play of the BOS treatments. Recall that the BOS game contains elements of conflict in

Table 7-4. Battle of the Sexes Game: Last 11 Periods[1]

Game	outside option	(1, 2)	(2, 1)	(1, 1) or (2, 2)	Total
BOS	—	37 (22%)	31 (19%)	97 (59%)	165
BOS-300	33	0 (0%)	119 (90%)	13 (10%)	165
BOS-100	3	5 (3%)	102 (63%)	55 (34%)	165
BOS-1W	—	1 (1%)	158 (96%)	6 (4%)	165
BOS-2W	—	49 (30%)	47 (28%)	69 (42%)	165

[1] Numbers in the parentheses refer to proportions of play in the subgame.

addition to coordination. The first row of this table illustrates the coordination problems. Only 41% of the time did we observe ex post equilibrium; 59% of the time both players received a zero payoff.

Again there is evidence to support the hypothesis that players employ forward induction reasoning. When row rejects a relevant outside option of 300, play is very focused on row's preferred pure strategy outcome (2, 1). Over 90% of the time that the subgame is played, this outcome is achieved. Both players clearly understood the message: after rejecting the option, row played strategy 2 98% of the time while 92% of the time column played strategy 1. However, as with the coordination game discussed above, the outside option is taken too frequently (20%). Perhaps of significance is the fact that the acceptance of the outside option was relatively concentrated in a subset of the subjects. Three (out of 33) players were responsible for almost 40% of the selections of the 300 option.

While forward induction would appear to have helped players coordinate their actions to achieve ex post equilibrium, the third row of table 7-4 reveals that something else might explain this success. When the outside option was so low that it should have been irrelevant, in this case 100, play nevertheless became much more focused on the (2, 1) outcome. This option is accepted very infrequently but, after rejection, the (2, 1) equilibrium is reached 63% of the time as opposed to only 19% in the simple BOS game.[16] While this is not as strong a coordination effect as that observed upon rejection of the 300 outside option, results from BOS-100 surely differ from BOS. We are lead to conclude that much of the success of BOS-300 may be due to focal point effects unrelated to forward induction.[17]

Table 7-5. Battle of the Sexes Game: Mapping of Announcements to Actions in 1-way Communication Game

Announcement by Row	Actions: Row, Column				
	(1, 2)	(2, 1)	(1, 1)	(2, 2)	Total
0	1	4	0	3	8
1	0	0	0	0	0
2	0	154	1	2	157
Total	1	158	1	5	165

Table 7-6. Battle of the Sexes Game: Mapping of Announcements to Actions in 2-way Communication Game

Announcements Row, Column	Actions: Row, Column				
	(1, 2)	(2, 1)	(1, 1)	(2, 2)	Total
0, 0	0	2	1	6	9
0, 1	0	2	0	0	2
0, 2	21	0	0	6	27
1, 0	2	0	0	1	3
1, 1	0	0	1	0	1
1, 2	14	0	0	0	14
2, 0	0	24	3	3	30
2, 1	0	7	0	1	8
2, 2	12	12	4	43	71
Total	49	47	9	60	165

In contrast to the coordination games, here one-way cheap talk communication coordinates play on (2, 1) more frequently than either a relevant outside option or two-way communication. In fact one-way communication successfully coordinated play at (2, 1) 96% of the time. As table 7-5 reveals, this was accomplished by row announcing strategy 2 over 95% of the time (and choosing to be silent—strategy 0—the remaining periods). Over 98% of the time these announcements of 2 led to play of (2, 1) as predicted by Farrell.

Two-way communication was not as successful, leading to coordination at (2, 1) or (1, 2) only 58% of the time. Note, however, that this is much greater coordination than that achieved in BOS, so communication is helping. The announcement data in table 7-6 give some idea where coordination breaks down. On the 22 occasions that either (1, 2) or (2, 1) was announced, play conformed to announcements 21 times, consistent with Farrell's theory. Problems arose, in part, because 43% of the time both players announced (2, 2). And while choosing silence (0 in table 7–6) in general seemed to be an attempt to coordinate, this was not universally understood.[18]

This highlights the important differences between these games. In BOS the players' problem is one of choosing between two equally attractive equilibria about which they have conflicting rankings and neither strategy is less risky than the other. Players need a mechanism to make one of the equilibria focal and nothing does that more clearly than having one player announce his intended strategy. Two-way communication introduces the possibility of conflicting messages and forward induction is a slightly more subtle solution to the coordination problem.

In the coordination games, however, there is no question about the desired outcome. The problem here, given the strategic uncertainty and the presence of a safer strategy that does not support the best outcome, is for players to develop sufficient confidence that their rivals will play the strategy which supports the Pareto-dominant equilibrium. Among the institutions studied, this is best accomplished by allowing both players an opportunity to announce their intentions.

5. Conclusions

The point of these experiments has been to provide an understanding of the nature of coordination problems endemic to many economic interactions. In our experimental work on coordination, we have established that agents do not play the Pareto-dominant Nash equilibrium in a variety of Coordination Games and that ex post disequilibrium arises quite often in the Battle of the Sexes Game.

In light of these coordination difficulties, it is quite natural to consider alternative institutions (i.e., different extensive form games) that might overcome them. We summarized the results from two important variations on these games: the presence of an outside option and pre-play communication. These institutions are both relevant in that they provide a means of coordinating activities. Moreover, these same institutions have received considerable attention in the game theory literature.

Our findings indicate that neither preplay communication nor the presence of outside opportunities will necessarily solve all coordination problems. In particular, we find that the nature of the strategic interaction between players—coordination v. conflict—determines which institution fares better. Our work indicates that in games of conflict, such as the BOS game, one-way communication or an outside option for one agent is sufficient for resolving ex post disequilibrium problems because these institutions provide an asymmetry in the game. In fact, we find that even a payoff irrelevant outside option influences play through the creation of an asymmetry.

For coordination games, in contrast, the issue is confidence and the creation of an asymmetry is not relevant. For CG, we find that only through two-way communication is it possible to build enough confidence to attain the Pareto-dominant Nash equilibrium.

These results are suggestive along two lines. First, they point to the need to develop a crisper characterization of the differences between games of conflict and games of coordination. Second, using this characterization, one should continue to test the theme of this paper regarding the nature of the strategic interaction and the choice of institution for resolving the particular coordination problem.

Notes

1. This is the spirit of Schelling's (1960, 1978) original discussion of coordination games in which players are indifferent between all equilibria. In Schelling's coordination games, the concern was one of avoiding ex post disequilibria.

2. See Cooper et al. (1989) (hereafter CDFR2), (1992a) (CDFR3), (1991) (CDFR4) and (1992b) (CDFR5).

3. That is, strategy 1 is risk dominant in the sense of Harsanyi and Selten [1988].

4. Examples of BOS-type games have also appeared recently in the theoretical literature, most notably with reference to problems of entry in naturally monopolistic industries (Farrell, 1987, and Dixit and Shapiro, 1986) and of product standardization (Farrell and Saloner, 1985). A BOS-type repeated game is at the core of a model that Bolton and Farrell, 1990, use to support centralized decision-making as a means to resolve coordination problems.

5. See for example, CDFR1, CDFR3 and Van Huyck et al., 1990, for versions of CG and CDFR2 for BOS.

6. Forward induction is a central feature of the Kohlberg and Mertens (1986) notion of strategic stability in normal form games. For an extensive discussion on the relationship between forward induction and strategic stability see van Damme (1989, 1990).

7. For the purposes of this discussion we have assumed that when players are permitted to communicate their intended action they cannot choose to be silent. In fact, in our earlier work on communication in the battle of the sexes game (CDFR2) we did permit players to choose silence and as the data reported below indicate, some did. To integrate this option into a theory of cheap talk requires an assumption of the meaning of silence. In CDFR2 we argued that silence should not ever be chosen in one-way communication games since the sender can choose his preferred equilibrium by announcing 2. In two-way communication, however, there may be a role for silence and we argue that it is used as a coordinating device; e.g. row's choice to be silent is understood as an agreement to go along with whatever equilibrium column's announcement suggested.

8. See Roth-Malouf (1979) and Berg et al. (1986) for a complete description of the procedure.

9. All Session I games had $1.00 prizes while Session II games had prizes of $2.00 for BOS and $1.00 for CG.

10. In both sessions, player pairings followed an eleven period sequence in which each player played every other player exactly once and each alternated between being the row and column player during the periods in which he was participating. These pair assignments were randomized at the beginning of each eleven periods to prevent players from playing against the same opponent in the same order within each eleven period sequence. Players were told that they would play every other player once in Session I and twice in Session II (once as a row player and once as a column player). They were not told anything more about the sequence of matches in Session II.

11. We wanted the pairings to satisfy two conditions: (i) players were to alternate being row and column players, and (ii) each player was to play each of the other players once (in Session I) or twice (in Session II). It is impossible to satisfy these two conditions with an even number of players. The player who sat out in each period drew the lottery ticket which determined the prize. Having the player who sits out draw the lottery ticket may serve the additional purpose of helping to convince players that the lottery was run fairly.

12. The ability of players to recognize dominated strategies is particularly important in these games since the logic of forward induction requires the iterated deletion of dominated strategies.

13. Each cohort completed the two sessions in about 1 and 1/2 hours. Payments to participants ranged from $6 to $33.

14. These results also appear in our previously published papers: BOS, BOS-1W and BOS-2W in CDFR2; BOS-300 and BOS-100 in CDFR4; and, CG, CG-900 and CG-700 in CDFR5.

15. We do not focus on the earlier periods of play as we are not investigating any theories concerning the process of reaching an equilibrium.

16. Row seemed to expect (2, 1) more than column. Conditional on rejecting the 100 outside option, row played strategy 1 only 13% of the time while strategy 2 was played by column 27% of the time.

17. In CDFR4 we reported on a related treatment in which game BOS was played with the following difference: the row player moved first, but his/her move was not revealed to column until after column had selected his/her action. Strategically equivalent to the simultaneous move game BOS, play nevertheless was different; in fact, very close to that in BOS-100. Players seemed to be using the known sequence of moves to define a focal equilibrium. This is consistent with the MAPNASH refinement proposed by Amershi, Sadanand and Sadanand [1989a, b].

18. From table 6, we can see that announcements of (0, 1), (1, 0), (0, 2) (2, 0) led to the play of (2, 1), (1, 2), (1, 2) and (2, 1) respectively, 49 out of 62 times (79%). In addition, the fact that the silence option was much more popular in the two-way treatment also suggests that players may have been using it to coordinate.

References

Amershi, A., A. Sadanand, and V. Sadanand. 1989a. "Manipulated Nash Equilibria I: Forward Induction and Thought Process Dynamics in Extensive Form." Discussion Paper 1989–4, August.

Amershi, A., A. Sadanand, and V. Sadanand. 1989b. "Manipulated Nash Equilibria II: Applications and a Preliminary Experiment." Discussion Paper 1989-6, August, University of Minnesota.

Berg, J.E., L.A. Daley, J.W. Dickhaut, and J.R. O'Brien. 1986. "Controlling Preferences for Lotteries on Units of Experimental Exchange." *Quarterly Journal of Economics* 101: 281-306.

Bolton, P., and J. Farrell. 1990. "Decentralization, Duplication and Delay." *Journal of Political Economy* 98: 803-826.

Bryant, J. 1983. "A Simple Rational Expectations Keynes-Type Model." *Quarterly Journal of Economics* 98: 525-529.

Cooper, R., D.V. DeJong, R. Forsythe, and T.W. Ross. 1990. "Selection Criteria in Coordination Games." *American Economic Review* 80: 218-33. (CDFR1)

Cooper, R., D.V. DeJong, R. Forsythe, and T.W. Ross. 1989. "Communication in the Battle of the Sexes Game." *Rand Journal of Economics* 20: 568-587. (CDFR2)

Cooper, R., D.V. DeJong, R. Forsythe, and T.W. Ross. 1992a. Communication in Coordination Game." *Quarterly Journal of Economics* 107: 739-771. (CDFR3)

Cooper, R., D.V. DeJong, R. Forsythe, and T.W. Ross. 1991. "Forward Induction in the Battle of the Sexes Game." Working paper, University of Iowa, forthcoming, *American Economic Review*. (CDFR4)

Cooper, R., D.V. DeJong, R. Forsythe, and T.W. Ross. 1992b. "Forward Induction in Coordination Games." *Economics Letters* 40: 167-172. (CDFR5)

Cooper, R., and A. John. 1988. "Coordinating Coordination Failures in Keynesian Models." *Quarterly Journal of Economics* 103: 441-463.

Cooper, R., and T. Ross. 1985. "Product Warranties and Double Moral Hazard." *Rand Journal of Economics* 16: 103-113.

Diamond, P. 1988. "Aggregate Demand Management in Search Equilibrium." *Journal of Public Economics* 90: 881-894.

Dixit, A., and C. Shapiro. 1986. "Entry Dynamics with Mixed Strategies." In L.G. Thomas (ed.), *The Economics of Strategic Planning*. Lexington: Lexington Books.

Farrell, J. 1987. "Cheap talk, Coordination, and Entry." *Rand Journal of Economics* 18: 34-39.

Farrell, J., and G. Saloner. 1985. "Standardization, Compatibility, and Innovation." *Rand Journal of Economics* 16: 70-83.

Harsanyi, J, and R. Selten. 1988. *A General Theory of Equilibrium Selection in Games*. Cambridge, Mass.: MIT Press.

Heller, W. 1986. "Coordination Failure under Complete Markets with Applications to Effective Demand." In *Equilibrium Analysis, Essays in Honor of Kenneth J. Arrow, Volumn II*, Walter Heller, Ross Starr, and David Starrett (eds.). Cambridge: Cambridge University Press.

Katz, M., and C. Shapiro. 1985. "Network Externalities, Competition and Compatibility." *American Economic Review* 75: 424-440.

Kohlberg, E., and J. Mertens. 1986. "On the Strategic Stability of Equilibria." *Econometrica* 54: 1003-1038.

Roth, A.E., and M.W.K. Malouf. 1979. "Game-Theoretic Models and the Role of Bargaining." *Psychological Review* 86: 574–594.
Schelling, T. 1960. *The Strategy of Conflict.* Cambridge: Harvard University Press.
Schelling, T. 1978. *Micromotives and Macrobehavior.* New York: Norton & Company.
Smith, Maynard J. 1982. *Evolution and the Theory of Games.* Cambridge: Cambridge University Press.
Van Damme, E. 1987. *Stability and Perfection of Nash Equilibria.* Heidelberg: Springer-Verlag.
Van Damme, E. 1989. "Stable Equilibria and Forward Induction." *Journal of Economic Theory* 48: 476–496.
Van Damme, E. 1990. "Refinements of Nash Equilibrium." Center for Economic Research, Tilburg University, The Netherlands.
Van Huyck, J., R.C. Battalio, and R.O. Beil. 1990. "Tacit Coordination Games, Strategic Uncertainty and Coordination Failure." *American Economic Review* 80: 234–248.

III COORDINATION IN SPECIFIC ECONOMIC CONTENTS

8 THE DYNAMICS OF BANDWAGONS

Joseph Farrell* and Carl Shapiro*

1. Introduction

The diffusion of a technological advance is seldom smooth. Typically, some users hesitate to adopt a new technology until others have done so. And even when a new technology is adopted, different early users often choose distinct and perhaps incompatible versions of the new technology—either because these users differ in some relevant way, or just because they see no compelling need to standardize and therefore almost accidentally diverge.

Such diversity can be desirable. It is often beneficial for many different versions of a new technology (developed or sponsored, perhaps, by different vendors) to coexist and compete at least for some time. But where compatibility is important, as in the information processing and communication industries, the benefits of compatibility may come to outweigh those

---

* Farrell thanks the Hoover Institution and the National Science Foundation for financial support. Shapiro thanks the National Science Foundation, the John M. Olin Foundation, and the Sloan Foundation for financial support. This paper was begun while Shapiro was a Fellow at the Center for Advanced Study in the Behavioral Sciences.

of diversity—save for the fact that many users are now "locked in" to their earlier choices.

For example, different companies use incompatible word processing systems or incompatible data base programs.[1] Cellular telephone operators have adopted incompatible digital technologies to replace analog systems.[2] Human languages and railroad gauges are (less high-tech) examples of the same phenomenon.

If compatibility benefits grow more important with time, users may eventually overcome historical lock-in and choose to bear the costs of switching to a single variant of the new technology that promises to become a standard. Because such a switch grows more attractive the more other users are making it, this process can exhibit positive feedback,[3] in which one version of a technology acquires more and more users, thus becoming more and more attractive and so attracting still more users. This is a *bandwagon*.

In this paper we use some simple models to explore the dynamics of such bandwagons. We begin with the simplest possible dynamic model of bandwagons, studying the process by which a *single, standard* version of a new technology is adopted. In this model, a number of user groups are initially using "old" technologies and must decide when to adopt the new technology. This model lets us address some issues of timing: who adopts when. We find that adoption often involves users with different preferences adopting at the same time, but this coordination occurs only up to a point: sufficiently diverse users may wait for the next wave of adoption. Adoption is generally too slow; this is unsurprising given the positive adoption externality.

Although such timing issues will appear in some form in all bandwagon models, the bandwagon process is generally more complicated than our first model suggests, because these is not a clearly identified new technology to which all movement takes place. Rather, bandwagon processes often constitute the market's only mechanism of "choice" among a range of candidate technologies. This raises a number of other questions; a particularly important one concerns how bandwagons deal with, or perhaps create, uncertainty about which technology will be the eventual winner. This question requires a model with two or more new technologies whose relative merits are unclear.

We leave that question for future work, and focus here on a related question: in a model without uncertainty, what determines not only the timing of adoptions but, more fundamentally, which technology will prevail? Compatibility is often achieved *ex post* by means of some groups adopting technologies compatible with others', despite some cost of doing so. Much of our paper, then, is devoted to this question: In such a framework, if users already have adopted incompatible versions of a new technology, which version or versions of this new technology will survive or attract a bandwagon?

We do not ask here how or why users originally picked incompatible systems. In the case of human languages and, perhaps, computer systems, early adopters probably saw little benefit from compatibility, or may even have been unaware of the issue. In some cases, incompatibility may have been deliberately chosen by vendors, who may have believed that it favored their market prospects.[4] Clearly, whatever the reasons, many technologies are not standardized in advance of their adoption, so the starting point for this analysis, namely the presence of installed bases for several incompatible systems, is perhaps even the norm. But for a variety of reasons the (recognized) importance of compatibility has grown strongly over time. This is the starting point of our models below.

Given initial incompatibility among the groups, along with a growing desire to achieve compatibility, what determines which system attracts users from other systems and thus forms a bandwagon? Will only one bandwagon start up, or will there be (at least for a time) "competing" bandwagons? If so, how do bandwagons compete? How quickly will a bandwagon grow? Which systems will be abandoned, and when? Which is more likely to attract a bandwagon: a high-quality system, or one that begins with a large installed base? What happens in the competition between a technology with a large installed base and one whose users are highly entrenched, i.e., one that is costly to switch away from? These are the questions we address below.

We provide necessary and sufficient conditions under which multiple simultaneous adoptions of a new technology, "avalanches," occur in equilibrium. In competition between two technologies to become bandwagons, we show that the market equilibrium is biased in favor of the technology with the larger initial installed and against the technology offering higher quality.

Finally, we turn to a horizontally differentiated model, in which moving costs depend on "how far" a group moves. We show that aspects of behavior such as the degree of avalanching depend on the form of the moving-cost function.

2. Elementary Bandwagon Dynamics

2.1. A Model of Technology Adoption

We begin with a dynamic model of a *single* bandwagon. There are N agents initially using various "old" technologies. These agents should be interpreted as groups of users who are *administratively coordinated*. We use the terms "group" and "agent" interchangeably. We suppose that decision-making for each group is centralized, or equivalently that all users in a group have identical interests (and communicate efficiently), so a group always sticks together. For example, a firm or a division within a firm may select a database system for all of its employees to use, and cellular-telephone companies select systems that will be used throughout their regions of coverage.

The N groups will in general differ in size. We write s_i for group i's size, i.e., the number of users in group i.[5] We denote by $S \equiv \Sigma_i s_i$ the total number of users in all groups. The flow payoff to each member of group i when it is using the old technology is given by b_i. These b_i may differ for many reasons, including the vintage of a group's installed capital, the particular version of older technology used by a given group, and especially the size of the group: if network externalities are important in the old as well as the new technology, a larger group is likely to have a higher stand-alone payoff flow.[6]

Each group has but a single decision: when, if ever, to adopt the new technology.[7] Formally, then, we are studying a game of timing in which the payoffs are highly non-convex: each agent's benefit of moving is greater, the more agents are moving as well.

Adoption by group i involves a one-time, per-capita cost of $c_i > 0$. These *switching costs* or *moving costs* or *adoption costs* might include expenditures on new capital equipment, retraining to use the new technology, the costs of transferring existing information (e.g., files or macros) to the new system, etc. We permit these switching costs to vary across users, although we make the strong assumption that group i's moving cost c_i is independent of time.

The benefit of adoption by group i is that the members of the group can begin to enjoy the flow benefits of the new technology. This technology is subject to network externalities, i.e., the users value compatibility with each other; thus the flow benefits at any date depend upon the total number of users who have adopted the new technology by that date. We denote the flow payoff to an individual using the new technology at time t by $q(t) + \beta(t)v(S(t))$. Here the nondecreasing function $q(t)$ represents the *quality* of the new technology at date t,[8] $S(t)$ is the total number of individuals—that is, the sum of the sizes of the groups—using the new technology at time t, the nondecreasing function $v(\cdot)$ is the *network benefit function*, and $\beta(t)$ measures the value of compatibility at date t. Group i's overall payoff is the present value of the flow of benefits its members receive, less any moving costs they incur. All groups discount future benefits and costs using the interest rate r.

We assume that compatibility is growing more valuable over time: $\beta'(t)$ > 0 at all t. This increasing value of compatibility, and (less centrally) any increases in quality over time, are the exogenous forces in our model pushing the agents towards standardization on the new technology.

Formally, we analyze the following dynamic game in continuous time. At each instant, each agent can either switch to the new technology or simply continue to use the old. At any time, it is common knowledge which agents have adopted the new technology.[9] What is the timing of adoption of the new technology?

We use the equilibrium concept of Markov Perfect Equilibrium to answer this question. Two technical matters deserve mention here. First, because it is the growing flow benefits that drive the dynamics, each group i separately wishes to adopt at the time when the flow benefits just balance the flow cost rc_i of adopting earlier: it need not consider the rate of growth of benefits, for example. If moving costs were declining over time, agents would need at least local foresight.[10]

Second, it is well known that, when network externalities are important, there may be a range of adoption equilibria: roughly speaking, a time T can be chosen arbitrarily within some limits and it will be an equilibrium for everyone (or many agents) to adopt exactly at T.[11] Throughout this paper, we adopt a refinement that, in such multiple-equilibrium problems, picks out the earliest adoption equilibrium. Our reason for doing this is that the adoption game is supermodular, so an earlier equilibrium is Pareto-preferred to a later equilibrium; moreover, there is nothing to be gained from fooling others into adopting too early. Our refinement therefore makes sense if agents can talk to one another, as is common in the applications we have in mind.

2.2. Equilibrium Bandwagon Behavior

We can completely characterize equilibrium bandwagon behavior in this simple model. Rank the agents according to $rc_i + b_i$. Specifically, label the groups so that $rc_1 + b_1 \leq rc_2 + b_2 \leq \cdots \leq rc_N + b_N$. With this convention, define $S_i \equiv s_1 + \cdots + s_i$.

Lemma. *Group j cannot adopt before group i if $j > i$.*

Proof. See the Appendix.

This lemma, importantly, tells us the order in which the groups adopt. We next characterize the times of adoption and, more importantly, the qualitative dynamics: whether multiple groups adopt the new technology simultaneously in adoption "avalanches" or "cascades," or whether generically only one group adopts at a time (as would be the case absent network externalities).

To do this, form the ratios:

$$\theta_i \equiv \frac{rc_i + b_i}{v(S_i)}. \tag{1}$$

Both the numerators and the denominators of the θ_i's increase with i, so in general θ_i is not monotonic in i. We will show that the order of adoption is determined by a combination of the ordering by moving costs ($rc_i + b_i$) and the ordering by θ_i.

Proposition 1. *Suppose $\theta_k = \min_i \theta_i$. The initial adoption of the new technology takes the form of an avalanche by groups 1 through k and occurs at the date T_k^* defined by $\beta(T_k^*) = \theta_k$.*

Proof. See the Appendix. ∎

The essentials of the proof are not complex: No group would possibly want to move before T_k^*, but movement at that time is attractive given the coordination among groups $1, \cdots, k$. Recall our assumption that Pareto-superior outcomes involving earlier adoption are achieved by communication and coordination among groups.

Repeated application of Proposition 1 in fact characterizes the entire path of adoption of the new technology. After groups 1 through k adopt, suppose $\theta_m = \min_{i>k} \theta_i$. Then the next adoption date is T_m^* and involves groups $k+1, \cdots, m$. This same procedure is then repeated until all groups have adopted, or until only groups $n, \cdots, N$ remain and $\theta_i > \lim_{t \to \infty} \beta(t)$ for all $i \geq n$, in which case groups $n, \cdots, N$ never adopt the new technology.

Figure 8–1 illustrates this procedure. As shown groups 1 through m move first, then there are lone movements by two groups; finally all remaining groups move at once.

2.3. Timing of Adoption

As one might expect, equilibrium adoption in the presence of network externalities is not efficient. In particular, since adoption confers positive externalities on all others who have already adopted, market adoption is generally later than would be efficient in this model.[12] More precisely, if

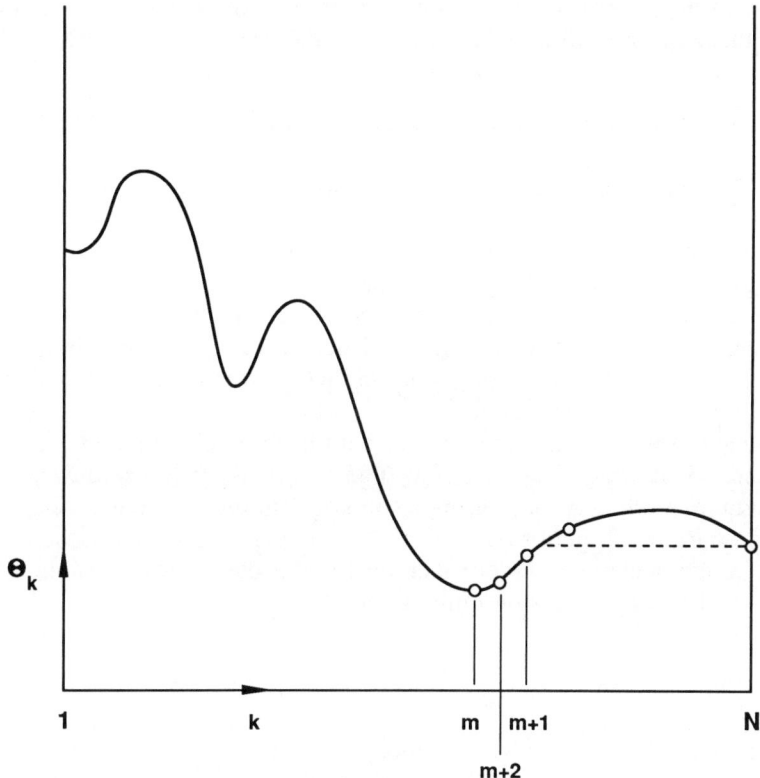

Figure 8–1. Path of adoption of new technology

groups $i, \cdots, j$ adopt at the same time in equilibrium, then by Proposition 1 they adopt at the time that group j prefers (given that they all adopt together). Earlier adoption by these groups would be preferred by groups $i, i+1, \cdots, j-1$, as well as by any groups already using the new technology. Only if *all* groups have the same $rc_i + b_i$, so they all agree on their preferred adoption time, is equilibrium efficient. In this case, equilibrium involves a complete avalanche. Generically, however, the $rc_i + b_i$'s differ so equilibrium is not efficient.[13]

The finding that equilibrium adoption is generically too late, even given our assumption of communication and coordination among the agents, should serve as a reminder that coordination is especially important in network industries. Even a slight breakdown in coordination that causes a delay in adoption will exacerbate a preexisting market imperfection and will therefore have first-order adverse welfare effects.

3. A Dynamic Model of Bandwagon Rivalry

For the remainder of the paper we study the situation in which, rather than forming a bandwagon on a single new technology that is new to all groups, the N agents initially are using incompatible versions of a *new* technology (incompatible "technologies" or "systems"), and compatibility is achieved by means of some groups switching, despite a cost, to others' systems. Each agent therefore has a richer set of choices than in the simpler model above: it must pick not only *when* to move (if at all) but also *which* other system to adopt.[14]

Again, agents value compatibility with each other. The flow payoff to an individual using technology i at time t is $\beta(t)v(S_i(t))$, where $S_i(t)$ is the total number of individuals using technology i at time t.[15] Again we employ Markov Perfect Equilibrium as our solution concept, with our assumption that communication and coordination lead to the earliest possible adoption in situations involving multiple equilibria.

3.1. Static Analysis

As a preliminary to our dynamic model consider briefly a static game with the basic structure outlined above. That is, fix t, and thus $\beta(t)$, and consider the one-shot game in which each agent can either move (at a cost) to join another agent or stay put.[16] If network benefits are unimportant, (β is small), the unique equilibrium of this static game is for all agents to stay with their own system. But is network benefits are important (β is large), there are multiple equilibria in the static game. Each agent wants to stay put if it believes others will move to join it, but is prepared to join a bandwagon around any other agent. This plethora of equilibria is both well known and commonplace in situations of network externalities. A static analysis, moreover, cannot answer questions about the dynamics of coalitions and standards.

3.2. Dynamic Analysis

Instead, our dynamic approach both narrows the set of equilibria and permits us to study questions of timing. As above, we resolve the multiple-equilibrium problem of "who moves?" by assuming that the group that is first willing to move does so. In much of our analysis this is, as we shall show, equivalent to endorsing a natural equilibrium selection for the case of two groups ($N = 2$) and pursuing (by induction) its implications with more general N.

4. Two Groups

The special case of two groups highlights some issues that arise with an arbitrary number of groups, and is a necessary starting point since our equilibrium refinement assumption is formally made with just two groups and extended by induction to many groups.

4.1. Equilibrium with Two Groups

With two groups, our dynamic model is a waiting game. At any instant, each group would prefer that the other groups switch rather than switching itself; and once a group switches, nothing further happens.[17]

If group 2 expects that group 1 never will switch, then group 2 faces a simple decision problem: What is the optimal date to switch, if ever? Given that $\beta(t)$ is increasing over time, group 2's optimal switching time T_2^* is characterized by the condition that the added flow payoffs from switching earlier just equal the extra interest cost from incurring the switching cost c_2 earlier:

$$\beta(T_2^*)(v(s_1 + s_2) - v(s_2)) = rc_2. \quad (2)$$

For example, if $\beta(t) \equiv bt$ and $v(s) \equiv ws$ then the time at which group 2 would be willing to move[18] is $T_2^* = (r/bw)(c_2/s_1)$; intuitively, group 2 gains network benefits proportional to s_1 per member and incurs a cost per member of c_2. Similarly, if group 1 believes that group 2 will never move, then its optimal moving date T_1^* is given by

$$\beta(T_1^*)(v(s_1 + s_2) - v(s_1)) = rc_1. \quad (3)$$

What does this tell us about equilibrium in the game? One obvious candidate equilibrium is for group 2 to adopt group 1's system at date T_2^*. Of course, we must specify out-of-equilibrium beliefs that do not tempt group 2 to delay in the hope that 1 will then move: for example, we could specify that group 1 believes that group 2 will move at all dates including and after T_2^*, and group 2 believes that group 1 will never move. A second candidate equilibrium, of course, is for groups 1 to move at T_1^*, with the analogous supporting beliefs.

When are these indeed equilibria? We will show that one always is, and the other may be. Temporarily define the groups so that $T_2^* \leq T_1^*$. Then it is certainly an equilibrium for group 2 to join group 1 at date T_2^*, with the system of beliefs indicated above. Group 1 cannot hope to earn a higher payoff than it gets in this equilibrium (since group 2 would never move before T_2^*), and group 2's beliefs that group 1 will not move, even after T_1^*, deter group 2 from deviating and are consistent in equilibrium.

Is it also an equilibrium for group 1 to join group 2 at date T_1^*? It is, if this outcome gives group 2 a payoff at least as large as the previous equilibrium. In that case, group 2 is willing to wait until T_1^*, since it believes that it can thus avoid having to switch. (If group 2's payoff from waiting were less than in the first equilibrium, however, it could profitably deviate by simply moving at T_2^*.) The condition for this second equilibrium to exist is, therefore,

$$\int_{T_2^*}^{T_1^*} (v(s_1 + s_2) - v(s_2))\beta(t)e^{-r(t-T_2^*)}\,dt \leq c_2. \tag{4}$$

There are no other pure-strategy equilibria. Given that group i moves, it prefers to move at time T_i^*. Therefore, group i cannot move earlier than T_i^* with any positive probability; it is better off waiting until T_i^* to move, and better off yet if the other group moves to it in the interim. Nor can group i (with probability 1) move after T_i^* in equilibrium, for it would choose to move earlier instead.[19]

Our basic equilibrium refinement is the following. In all games involving two groups (including two-group subgames of larger games) we assume that the "focal" equilibrium (the equilibrium that is played) involves group 2 moving to join group 1 at date T_2^*. This is the pure-strategy equilibrium that always exists.[20] The equilibrium involving later movement, when it exists, is rather perverse: it calls for group 2, the group that is more eager to move, to wait and then "convince" group 1 that group 1 must move if compatibility is to be achieved. We call this perverse even though it is favored by the (controversial) forward-induction refinement.[21]

4.2. Who Switches in Equilibrium?

What determines which group in fact switches in equilibrium? By our analysis and assumption above, this is equivalent to the question: which group would choose to switch sooner if it has no hope that the other will switch? That is, which is smaller: T_1^* or T_2^*?

Clearly, $T_2^* < T_1^*$ if and only if

$$\frac{v(s_1 + s_2) - v(s_2)}{v(s_1 + s_2) - v(s_1)} > \frac{c_2}{c_1}, \tag{5}$$

which we can rewrite as

$$\frac{v(s_1 + s_2) - v(s_2)}{c_2} > \frac{v(s_1 + s_2) - v(s_1)}{c_1}. \tag{6}$$

Group 2 switches in equilibrium if and only if its ratio of incremental benefit (from compatibility) to switching cost is larger than is group 1's.

4.3. Efficiency

Is this equilibrium outcome efficient? We can decompose efficiency into two parts: 1) Does movement occur at the efficient time? and 2) Does the group with lower total moving costs switch? The first part is straightforward: Since movement generates positive external benefits, equilibrium movement is inefficiently late, given which group switches. We therefore focus on the second part. To abstract away from these straightforward issues of the timing of movement, we hold fixed the time of movement and compare welfare from group 1 moving with that from group 2 moving. (The comparison is independent of what time that is.) Given the date of coalescence, efficiency is simply a matter of minimizing total moving costs: the group i with the smaller value of $s_i c_i$ should move.

Label the groups now so that group 2 is the smaller; $s_2 < s_1$. Equation 5 tells us that group 2 moves in equilibrium if and only if

$$\frac{v(s_1 + s_2) - v(s_2)}{v(s_1 + s_2) - v(s_1)} > \frac{c_2}{c_1}. \tag{7}$$

But efficiency calls for group 2 to move if and only if

$$\frac{s_1}{s_2} > \frac{c_2}{c_1}. \tag{8}$$

For a linear network benefit function, $v(N) \equiv vN$, the left-hand side of (7) is the same as the left-hand side of (8), so the equilibrium outcome is efficient. But there is no reason to expect a linear $v(\cdot)$ in general: more plausibly $v(\cdot)$ is concave. As figure 8–2 shows, with a concave $v(\cdot)$ and with $s_2 < s_1$,

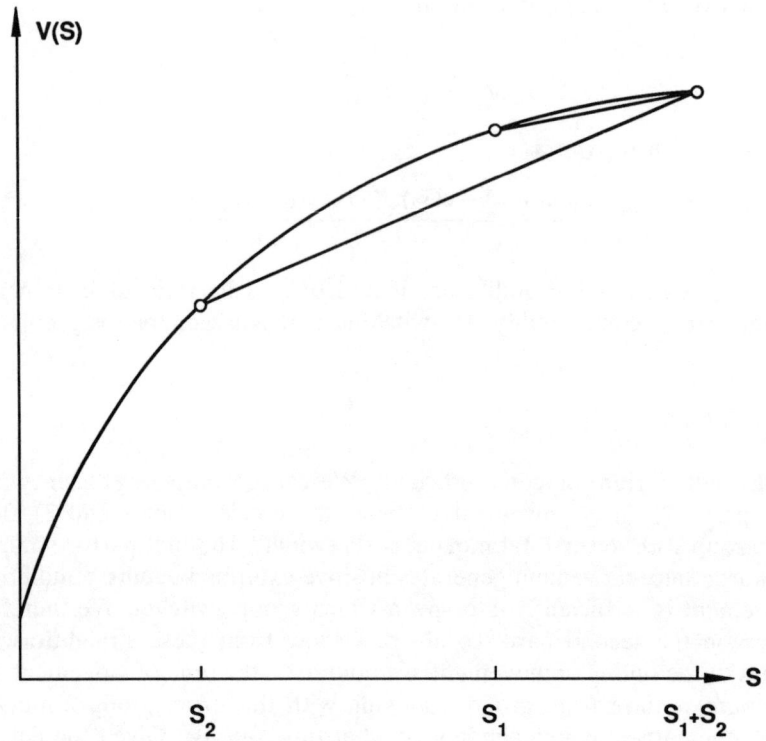

Figure 8-2. Bias in movement

$$\frac{v(s_1 + s_2) - v(s_2)}{s_1} > \frac{v(s_1 + s_2) - v(s_1)}{s_2},$$

which we can rewrite as

$$\frac{v(s_1 + s_2) - v(s_2)}{v(s_1 + s_2) - v(s_1)} > \frac{s_1}{s_2}. \tag{9}$$

From (9), we see that (8) implies (7), but not the converse. We record this as:

Proposition 2. *With two groups that may differ in size and per-capita moving costs, there is a bias against movement by the larger group if the* *network-benefit function* $v(\cdot)$ *is concave; it if is convex (in the relevant range) then the bias is reversed. If* $v(\cdot)$ *is linear, there is no bias.*

To understand this bias in favor of currently popular technologies, consider the relative size of the externalities conferred when either group moves. When group 1 moves, the total (flow) external benefit to group 2 is $s_2(v(s_1 + s_2) - v(s_2))$, which is equal to $s_1 s_2$ times an average slope of $v(\cdot)$ over the range $(s_2, s_1 + s_2)$. Likewise, when group 2 moves, the total externality is equal to $s_1 s_2$ times an average slope of $v(\cdot)$ over the range $(s_1, s_1 + s_2)$. With a concave function $v(\cdot)$, by equation (9), this externality is larger when the larger group moves; thus, equilibrium is biased against movement by the larger group. The bias comes about because of the concavity of the $v(\cdot)$ function.

4.4. Policy Implications

The efficiency analysis in the previous subsection does not explicitly discuss possible government policy in network industries that are moving towards de facto standardization. Yet it does identify a key externality in that process, and government often can play a crucial role in determining which of several competing technologies will prevail in a bandwagon process. For example, in computer networking, it seems that the US Government gave a crucial fillip to OSI standards over IBM's SNA system when it announced the phasing-in of a requirement that government procurement should conform to OSI.

One possible policy is for the government to decree a standard and compel all users to comply with that standard (immediately or, more probably, eventually). In our model, suppose that equilibrium involved group 2 switching to join group 1 at date T_2^*. The government could force the alternative outcome in which group 1 moves to join group 2. Absent additional policy instruments, however, group 1 would not move until the later date T_1^*.[22] Could such a policy approach be desirable? In other words, could efficiency be increased by insisting that group 1 move instead of group 2, even though this would involve later movement? The answer is given by the following Proposition:

Proposition 3. *With two competing technologies and a concave network-benefit function, it may be socially desirable to prevent users of the less popular technology from moving to join the more popular technology, but the reverse prohibition cannot be justified.*

Proof. Label the groups so that group 1 is larger, and write W_{ij} for the total welfare if group i moves at date T_j^*. If equilibrium involves movement by group 1, so $T_1^* < T_2^*$, our efficiency analysis above tells us that movement at any given date by group 1 is more efficient than movement at that date by group 2, so $W_{12} > W_{22}$. Furthermore, we know that $W_{11} > W_{12}$, since group 1 prefers to move at T_1^* rather than T_2^* and since earlier movement by group 1 benefits group 2 as well. We can conclude that $W_{11} > W_{22}$, so it cannot be socially desirable to prevent the large group 1 from moving.

Suppose instead that equilibrium involves movement by the smaller group 2, i.e., that $T_2^* < T_1^*$. Our efficiency analysis above tells us that equilibrium movement by group 2 may occur even if movement by group 1 at any given date T would be more efficient than movement by group 2 at T. If, as well, T_1^* is not much larger than T_2^*, overall efficiency will be increased by forcing the larger group 1 to move instead of group 2. ∎

The intuition behind Proposition 3 is straightforward. Any government prohibition of moving delays the formation of a bandwagon. This delay may be worthwhile if it corrects a bias towards the smaller group moving, but not otherwise.

4.5. Externalities from New Users' Adoption Choices

Government agencies may also affect technology choice in their role as users. In some cases, as in telecommunications, government agencies are large users. In other cases, government agencies may exert influence beyond their size, anointing a technology simply by choosing it.[23] How should government users of network technologies choose technology to promote overall efficiency? What externalities should such an agency consider in addition to its own interests? In this subsection we analyze the externalities from the adoption choice by a new user at time zero.[24] Specifically, suppose that a new user of size s_G will adopt technology 1 or 2 at time zero. The new adoption will increase s_1 or s_2 by s_G. Which choice will have more desirable external effects?

The choice has three effects on other parties. First, there is the *direct network effect* of enlarging one or the other network. On a flow basis, if the new user joins group 1, these direct benefits are given by $s_1(v(s_1 + s_G) - v(s_1))$. This flow must be integrated from time zero until the date at which groups 1 and 2 coalesce. For s_G small, the incremental flow benefits are approximated by $s_G s_1 v'(s_1)$. For small s_G, therefore, the externality is greater if the new user joins the group with the larger value of $s_i v'(s_i)$. The relationship of this quantity to the relative sizes of the groups is ambiguous: formally, the derivative of $sv'(s)$ is equal to $v'(s)(1 + sv''(s)/v'(s))$, so the sign of the relationship depends on the elasticity of v'. For example, if $v(s) \equiv s^\alpha$ for $\alpha > 0$, then $sv'(s)$ increases with s, so the direct effect argues for joining the initially larger group. But if $v(s)$ approaches a maximum at some level of s, and the larger group is already near this maximum, i.e., the large group has nearly exhausted the available network benefits, then there is little network externality generated if the new user joins the larger group, and the direct effect argues for the agency to join the smaller group.[25]

Second, the new user's choice will affect the date at which groups 1 and 2 come together. If equilibrium inevitably involves movement by group 2, joining group 1 will accelerate standardization, while joining group 2 will retard it. Clearly, this effect argues for joining the group that will eventually become the standard.

Third, the new user's choice *might* change the identity of the group that moves. To fix matters, suppose that in the absence of any new adoption group 2 would move to join group 1 at date T_2^*. The new user's choice of system might tip the process in one of three ways. a) The new user may be small but influential, so that its choice simply becomes focal. The analysis of this case is identical to that given above in which the government decrees a standard, except that the effect is limited to situations in which there are two equilibria to begin with. If technology 2 is initially less popular, it might be socially desirable for the new user to shift the bandwagon to technology 2. b) The new user might be large, so its choice of group 2 might cause T_2^* to exceed T_1^*. Again, this may generate greater external benefits than a choice of group 1 if group 2 is smaller than group 1. c) The new user may be small, but the choice of a standard may be "in the balance" in the sense that $T_1^* \approx T_2^*$.

We can explore this last case more fully by looking at the knife-edge case in which s_G is small, but $T_1^* = T_2^*$ so the choice by even a small new user will tip the balance to its chosen technology. Since s_G is small, the direct effect of the new user's choice is small, as is any acceleration effect. The tipping effect, however, need not be small. In this case Proposition 2 applies: assuming that $v(\cdot)$ is concave, the external effect is larger if the new user adopts the *less* popular technology.[26]

Proposition 4. *We have identified three external effects from a new user's choice of one competing technology over the other. The direct effect argues for choosing the technology with the higher value of $sv'(s)$; this may be the technology initially having the larger or the smaller installed base, depending* *upon the shape of the network benefit function v(·). The acceleration effect argues for picking the technology that is likely to win—that is, to be joined by the other group eventually. The tipping effect, if it applies, argues for joining the smaller installed base.*

5. Quality Differences

So far we have assumed that any two technologies are perfect substitutes if they have the same installed bases. We now relax that assumption by permitting the technologies to differ in quality. Specifically, we study competition between a technology with a large installed base and another that is of superior quality. Such competition is common, as when groups have previously made their technology adoption decisions at different times, thus embodying technologies of different vintages.

To examine this question, we now introduce quality into the agents' payoffs. Writing q_i for the quality of technology i, we assume that technology 1 is more popular ($s_1 > s_2$) but technology 2 is of higher quality ($q_2 > q_1$). To model quality, let the flow payoff at date t to an individual using technology i be $U(S_i, q_i, t)$. We assume that $U_{qt} = 0$, i.e., quality differences among the technologies are not changing over time. With this assumption, and retaining our assumption that network benefits grow over time, the flow payoff functions can be written (with a suitable calibration of quality) as

$$U(S_i, q_i, t) = q_i f(S_i) + \beta(t) v(S_i).$$

We assume that any interaction between quality and network size displays diminishing returns: $f' \geq 0$ and $f'' \leq 0$.

Our formulation includes several different versions of "quality." Network-independent quality is captured by the function $f(S_i) = k$, reflecting quality attributes that are independent of the size of the installed base. An example of network-independent quality might be the quality of the user interface of a piece of computer software; the value of a better (easier-to-use) interface may be independent of the size of the network. On the other hand, if $f' > 0$ the quality is "network-enhancing:" higher quality enhances the network externalities. An example of network-enhancing quality might be the speed of the modem in a personal computer.

Using the same line of analysis as we employed without the quality variable, the earliest date at which group 2 would move, T_2^*, is given implicitly by

$$q_1 f(s_1 + s_2) - q_2 f(s_2) + \beta(T_2^*)(v(s_1 + s_2) - v(s_2)) = rc_2.$$

This can be rewritten as

$$\beta(T_2^*)(v(s_1 + s_2) - v(s_2)) = rc_2 + q_2 f(s_2) - q_1 f(s_1 + s_2). \quad (10)$$

As above, group 2 moves in equilibrium if and only if $T_2^* < T_1^*$.

Equation 10 reveals that the analysis here of who moves and when is formally identical to that given above for two groups with no quality differences. We need simply replace the flow moving costs of rc_2 by the right-hand side of (10) to incorporate quality into our analysis. The new terms on the right-hand side of (10) capture the time-independent differences in flow benefits to members of group 2 when they switch to group 1. These in turn are based on differences in quality between the two technologies and the larger network that will prevail once group 2 has joined group 1.

The addition of the quality term $q_i f(S_i)$ into the benefit function reinforces our earlier finding that equilibrium is biased towards the smaller group moving to join the larger group. In other words, if the large group moves, this is efficient:

Proposition 5. *Suppose that, in competition between a popular technology 1 and a technically superior technology 2, equilibrium involves users of technology 1 switching to technology 2. Then, given the date of coalescence, movement by group 1 is more efficient than movement by group 1.*

Proof. The Proposition follows from Proposition 2 by formally replacing the moving cost with an adjusted moving cost reflecting quality differences. ∎

A less popular technology that is of superior quality (and/or has higher switching costs), and sufficiently so that it is efficient for all users to adopt that technology, may still fail to survive in the marketplace.

As an extreme case of this, suppose that there is just one established technology, technology 1, that has quality q_1 and on which there is an installed base of s_1 users. At a cost, they could switch to a newly developed superior technology of quality q_2; but the costs of switching outweigh (for them) the quality benefits of doing so, even assuming that they could coordinate the switch so that they would lose no compatibility benefits. Assume for definiteness that quality is network-independent. Then the effective per-capita switching cost for those users to switch to technology 2 is $c_1 - (q_2 - q_1)/r$. There are also s_2 uncommitted users. We can model the latter as being hypothetically "on" technology 2; the cost to each of these uncommitted users of "moving" to technology 1 is simply $(q_2 - q_1)/r$, the present value of the quality benefits sacrificed. Note that this does not include any "ordinary" cost of switching, but consists entirely of quality loss. This is appropriate because we do not think of the new technology as having an installed base.

Then the uncommitted users (group 2) will switch to 1 (i.e., will adopt the old, established technology) if and only if

$$\frac{v(s_1 + s_2) - v(s_2)}{v(s_1 + s_2) - v(s_1)} > \frac{q_2 - q_1}{rc_1 - (q_2 - q_1)}. \tag{11}$$

But efficiency calls for group 2 to move if and only if

$$\frac{s_1}{s_2} > \frac{q_2 - q_1}{rc_1 - (q_2 - q_1)}. \tag{12}$$

We can conclude that the market exhibits *excess inertia*: it is too likely to stick with an established technology when a superior new technology becomes available.

This result contrasts with previous analyses of the possibility of excess inertia. Farrell and Saloner (1985, 1986) show that either excess inertia or its opposite, insufficient friction, can occur. Katz and Shapiro (1986a, b) find a tendency towards insufficient friction when technologies are sponsored. In those models, there are two countervailing externalities: on the one hand, users may "strand" an installed base by adopting a new technology, thus stranding the installed base; on the other hand, users may harm future arrivals by adopting the old technology, thus effectively making it unavailable later. For convenience, those models also assumed that the installed base would never switch to the new technology. Here, by contrast, there are no future arrivals, and the installed base can switch to the new technology (although at a cost).

6. Coalition Formation with Equal Switching Costs and General N

We now extend our analysis by considering an arbitrary number N of groups. In order to keep the analysis tractable, however, we assume that all agents have the same quality (thus we suppress notation for quality) and the same per-capita moving costs: $c_i = c$ for all i. Groups differ (ex ante) only in their sizes s_i. For simplicity, we assume that there is a unique largest group; it should be clear how to modify our analysis if not.

6.1. Characterization of Equilibrium

The equilibrium pattern of movement is qualitatively similar to that found above in Proposition 1. Now, however, small groups move first (not those with low moving costs or poor baseline technologies).

For some period of time (possibly null, possibly forever), all groups stick with their original technologies. then some of the smallest groups (possibly just one, possibly all but the very largest) adopt the technology with the largest initial installed base (groups 1's technology). More time passes. Then some subset of the smallest remaining groups adopts group 1's technology. This process continues until either all groups have adopted technology 1 or movement ceases with some of the largest initial groups never moving.

Formally, order the agents so that $s_1 > s_2 \geq \cdots \geq s_N$. Write $\mathbf{s} \equiv (s_1, \cdots, s_N)$ for the "configuration of the system." We now form a new set of ratios:

$$\psi_i \equiv \frac{rc}{v(s_1 + s_i + s_{i+1} + \cdots + s_N) - v(s_i)} \tag{13}$$

We will show that the order of adoption is determined by a combination of the ordering by size and the ordering by ψ. Now we (re)define T_i^* by

$$\beta(T_i^*) = \psi_i. \tag{14}$$

Then group i will choose to move at date T_i^* if it will be accompanied (or preceded) by all smaller groups and by no others. Because their (per-capita) flow of benefits after all those groups move is the same as group i's, their moving costs are the same, and their opportunity-cost flow of benefits is smaller than i's, it follows that all the smaller members are all the more willing to do so. Thus, at time T_i^*, group i will move to group 1, and all smaller groups will move to group 1 if they have not already done so. If $T_i^* < T_j^*$ for $j < i$, then this movement will take place before any larger groups are ready to move.

As above, we assume here that the coordination required for all of these groups to move at date T_i^* is achieved, since moving is Pareto preferred by these groups to waiting and moving later. Such coordination is very plausible in standards-choice problems with a smallish number of agents: communication about who is planning to do what is both credible and common. Notice also that the date of this coordinated move is the *latest* of the participants' respective preferred dates. Thus movement is too late from almost all of the participants' point of view, as well as (obviously) from the target's point of view.

Based on this argument the same geometric procedure shown in figure 8–1 again can be used to determine who will move when, relying of course on the ψ_i's instead of the θ_i's. A very similar constructive procedure to the one used above again characterizes equilibrium: First plot the ψ_i's against i, as is shown in figure 8–1 for the θ_i's. Next, find the global minimum, say ψ_k. Then the first movement is at date T_k^*, by groups $k, k+1, \cdots, N$, i.e., by group k and all smaller groups, but by no bigger groups. Next, find the minimum of T_i^* for $i < k$. Say that is T_m^*. (Often m will be $k-1$.) The next movement is by group m and all smaller groups that have not already moved, i.e., by group(s) $m, \cdots, k-1$. And so on. Of course, if $\beta(\cdot)$ is bounded, there may come a time when not all groups have moved to technology 1 and yet no more movement occurs. Partial standardization may be the long run equilibrium outcome.

Proposition 6. *There is a unique subgame-perfect Markov equilibrium. All movement is to the biggest group. At any time, movement is by the smallest group(s) that have not yet moved, and it takes place at time T_k^* defined in equation (14) if group k is the largest group moving. Moreover, movement is faster if the biggest group is larger at the expense of smaller groups.*

Proof. See at Appendix. ∎

It is not difficult to write down conditions for there to exist an internal minimum of the T_i^* sequence; this is the condition for a "slide" (not a dribble, not a total avalanche). It is also easy to write down conditions for a total avalanche. With concave $v(\cdot)$, however, specific functional forms and parameter choices are necessary to reach further conclusions about the qualitative dynamics of adoption.

6.2. Linear or Convex Network Benefits

The equilibrium bandwagon dynamics take a particularly stark form if the network benefit function $v(\cdot)$ is linear or convex. In that case, equation (14) implies that the earliest T_i^* is T_2^*, which in turn implies that all groups move at the same instant to join the largest group, group 1. Thus we have:

Proposition 7. *Suppose that all groups have equal per capita moving costs and that network benefits are linear or convex in the number of individuals on the network. Then the unique equilibrium is for all agents except the* largest to move and join the largest agent; this movement takes place all at once, at the earliest date at which the second largest agent is willing to go along with such a bandwagon.

Note that here T_2^* depends only on $v(S) - v(s_2)$, so the distribution of individuals among groups $1, 3, 4, \cdots, n$ does not affect the equilibrium pattern or timing of movement.

Equilibrium also has a limited efficiency property with linear network benefits: the right groups move although (as usual) they move too late for full efficiency.

7. Bandwagons with Locations

So far, we have assumed that a group's moving costs are independent of the group it joins. Sometimes that assumption is appropriate; in other cases some moves will be more costly than others. Now we generalize our analysis by dropping that assumption and replacing it with a "location" structure. We formalize compatibility benefits by assuming that each group values being in the same "location" as other groups (i.e., compatibility), but we now put more structure on the cost of "moving," i.e., switching from one system to another. Allowing the cost of moving to vary with the "distance" moved allows us to study the dynamic market solution of the tradeoff between compatibility and variety when groups of users begin with their autarkically preferred technologies but, because network externalities are growing over time, eventually compromise in some way.[27]

Specifically, we assume, much as before, that group i enjoys network benefits from all other groups who are at *exactly* the same location as i.[28] To make up in part for these added complexities, we shall assume in this section that $v(\cdot)$ is linear. Thus, if we denote group i's location at date t by $x_i(t)$, the per-capita flow benefit to group i at date t is $\beta(t)S_i(t)$, where

$$S_i(t) \equiv \sum_{j \mid x_j(t) = x_i(t)} s_j,$$

i.e., $S_i(t)$ is the total number of users (including group i) compatible with i at time t.

From these gross flow benefits we must subtract i's *moving costs* to obtain i's payoff. We suppose that group i can move at most once, incurring a per-capita one-time moving cost c_i that varies with the distance moved, d. Specifically, we shall explore linear moving costs, $c_i \equiv \gamma_i d$, and quadratic moving costs, $c_i \equiv \gamma_i d^2$. Here γ_i is a group-specific moving-cost parameter that we will sometimes take to be 1.

7.1. Static Analysis and Myopic Dynamics

Our goal is to study the dynamics of who moves, when, and where. As a preliminary, however, consider a static model in which each group i (simultaneously) picks a new location $\hat{x}_i$, which may be equal or unequal to i's original location x_i. Then i's payoff depends, as described above, on the total size of the groups j, if any, that choose $\hat{x}_j = \hat{x}_i$, and on whether i moved ($\hat{x}_i \neq x_i$) and if so how far ($d_i \equiv |\hat{x}_i - x_i|$). In particular, with our special assumptions, each member of group i gets a flow payoff equal to $\beta S_i - rc_i$. This is a straightforward simultaneous-move game, expressing the idea that any agent, at a cost, can become compatible with another. If network externalities are unimportant (β is small), the unique equilibrium is that each group stays put: $\hat{x}_i = x_i$ for all i. If network size is important, other equilibria also exist; and for large enough β there will be many equilibria, as each group would be willing to move (to a wide variety of $\hat{x}_i$'s) in order to be compatible with other groups.

Thus the *static* analysis of the compatibility game is either trivial or fraught with multiple equilibria. Moreover, of course, being static it cannot answer questions about the dynamics of coalitions and standards. As we have seen above, by introducing dynamics in the form of $\beta(t)$ growing over time, along with an assumption that agents coordinate and move as soon as doing so is beneficial, we can avoid the multiplicity of equilibria in the static game and can discuss the timing of compatibility.

It would be satisfying to solve the dynamic game with locations using subgame-perfect equilibrium, but we have not been able to do this. Instead, we assume that groups are myopic in the following sense: in deciding whether or not to move (and where), a group is aware of others moving concurrently, but does not anticipate any greater willingness to move in the future by any other groups. Obviously this model does not assume as much foresight and "rationality" as subgame-perfect equilibrium, but it may be a more accurate description of behavior in some circumstances. With these expectations, our equilibrium concept involves movement as soon as a mutually profitable move becomes possible, in the sense that each group i is moving to some $\hat{x}_i$ at the optimal time for it to do so given that others are moving to $\hat{x}_i$.

Using this "myopic dynamics" methodology, we ask several questions in the following subsections about equilibrium movement to achieve compatibility:

1. *Compromise*: If two groups join, will they both move, or will one move all the way to join the other? Is the outcome efficient in this respect?
2. *Avalanches*: In the dynamic process, do many groups coalesce all at once, or is agglomeration gradual?
3. *Sensitivity*: Is the final outcome, or the path to it, discontinuously dependent on the initial parameters x_i, s_i, γ_i, etc.?
4. *Rationality*: Is the myopic-dynamic equilibrium also a subgame-perfect equilibrium? If not, what form does the irrationality take? For example, would groups do better to pay more attention to the need to move to large groups and pay less attention to proximity?

7.2. Compromise

How will two groups join—will one of them move all the way to the other, or will each move towards the other, i.e., will they compromise? More concretely, will one group adopt the other's standard, or will they jointly design a new standard that involves elements of each one's current practice? Each would like to persuade the other to adopt its standard or something close to it: in our model this saves on one's moving costs.

As we saw above, with moving costs that are independent of distance (i.e., without locations), efficiency calls for the group with the smaller $s_i c_i$ to move, and with a linear $v(\cdot)$, moving costs will indeed be minimized in equilibrium. In treating that model, we stressed inefficiencies with nonlinear $v(\cdot)$; now we will focus on inefficiencies that arise when $v(\cdot)$ is linear but moving costs depend on distance.

Call the distance between the two group $d_{ij} \equiv |x_i - x_j|$. With *linear moving costs*, group i is willing to move a distance d_i to join group j if and only if $r\gamma_i d_i \leq \beta(t) s_j$, and group j is willing to move by d_j to join i if and only if $r\gamma_j d_j \leq \beta(t) s_i$. Equilibrium movement occurs, by assumption, when it first becomes mutually beneficial. This date, t_{ij}, is given by

$$\beta(t_{ij}) = \frac{r\gamma_i \gamma_j d_{ij}}{s_i \gamma_i + s_i \gamma_j}.$$

At this time the groups will meet at their "center of gravity" (with each group weighted by its total moving cost, $s_i \gamma_i$):

$$x_{ij} = \frac{s_i \gamma_i x_i + s_j \gamma_j s_j}{s_i \gamma_i + s_j \gamma_j}.$$

With *quadratic moving costs*, efficiency does involve compromise, i.e., both groups should move. In fact, efficiency calls for the groups to meet at their ($s\gamma$-weighted) center of gravity x_{ij} defined above. Equilibrium also involves compromise: qualitatively, equilibrium is similar (in this respect) to the linear case. But equilibrium no longer involves movement to x_{ij}. Instead, it turns out that they compromise on something closer to the group with the *smaller* total moving costs, $s\gamma$.

To see this, consider the time and place at which movements is first mutually beneficial. If the two groups move then to location $\hat{x}_{ij}$, we must have

$$r\gamma_i(\hat{x}_{ij} - x_i)^2 = \beta(t)s_j,$$

and similarly

$$r\gamma_j(x_j - \hat{x}_{ij})^2 = \beta(t)s_i.$$

Hence,

$$\frac{(\hat{x}_{ij} - x_i)^2}{(x_j - \hat{x}_{ij})^2} = \frac{s_j\gamma_j}{s_i\gamma_i},$$

whereas

$$\frac{(x_{ij} - x_i)^2}{(x_j - x_{ij})^2} = \frac{(s_j\gamma_j)^2}{(s_i\gamma_i)^2}.$$

Thus, if $s_j\gamma_j > s_i\gamma_i$ and $x_i < x_j$, we find that $\hat{x}_{ij} < x_{ij}$.

We can conclude here, just as with linear moving costs, that the smaller or lower-moving-cost group does—in some sense—better than it should. This is just the opposite of our earlier results: namely that the large group was too inclined to sit tight and wait for the smaller. The explanation is that a shift in the meeting-point away from the equilibrium place towards the large group would cost *each member* of the smaller group the same amount as it benefits *each member* of the larger group; there is no mechanism to recognize the fact that there are more of the latter.

To summarize, we find that both the desirability of compromise, and the tendency to compromise, depend on the nature of moving costs. With fixed moving costs, there is no compromise in equilibrium, nor should there be, since compromise would require both groups to incur moving costs. With linear costs, there still should not be compromise, but in equilibrium there is: the group with the larger total moving costs in inefficiently forced to move. With quadratic moving costs, there should be compromise, and there is, but again the group with the larger total moving costs is badly treated.

7.3. Avalanches

Next, we ask whether several groups coalesce suddenly or gradually. A strong sense of "suddenly" is that all movement occurs simultaneously. A strong sense of "gradually" is that, generically, only two groups join together at any date t. One could imagine intermediate cases, but we do not find them.

To explore this question we assume that $\gamma_i = 1$ for all i: individuals do not differ in their cost of moving a given distance. Differences in the γ_i's will of course tend to produce more gradual dynamics, as different groups will resist moving to different extents.

We saw above that with linear network benefits and location-independent moving costs, there is a complete avalanche. By contrast, with linear moving costs, we conjecture that (generically) coalescence happens in pairs under our myopic dynamics. This means that, given N groups characterized by (x_i, s_i), the first pair can join together before the first triple, quadruple, ... can do so. In the Appendix we show, relying on numerical methods, that a pair can always form before any larger avalanche can form.

Proposition 8. *With three groups and linear moving costs, equilibrium (generically) involves a series of pairwise joinings with no larger avalanches.*

We conjecture that this result generalizes to any number of groups:

Conjecture. *With N groups and linear moving costs, equilibrium (generically) involves a series of pairwise joinings with no larger avalanches.*

Arguments supporting this conjecture are provided in the Appendix.

We thus conjecture in the linear-costs case that agglomeration happens gradually, in the sense that (generically) only two groups move at any one time. (Two groups that have joined have become one group, in this analysis.) This finding is in sharp contrast to the constant-costs model where (with linear $v(\cdot)$, as we assume here) a complete avalanche ensues.

With quadratic moving costs, we conjecture that an analogy to Proposition 8 will hold in that case as well: the attractiveness of the long-distance moving that is needed for avalanches is even less than in the linear specification. Indeed, we expect more generally that the tendency towards avalanches will depend upon the shape of the moving-cost function, $c(\cdot)$ and the shape of the network benefit function $v(\cdot)$. The increasing costs of large moves versus the increasing benefits of large coalitions should determine the size of avalanches in general.

7.4. Sensitivity

To what extent does the final outcome, or the path to the final outcome, depend "very sensitively" on the initial conditions x_i, s_i? The answer again depends on the shape of the moving-cost function.

Without locations, all groups join up at x_1 at the date given implicitly by $\beta(t)(S - s_2) = rc$. The final outcome therefore depends on the initial conditions only to the extent that the identity (location) of that initial largest group changes. Moreover, the path to this final outcome does not depend much on initial conditions: all groups move simultaneously, at a time that depends continuously on $S - s_2$.

With linear moving-costs, every coalescence preserves the system's center of mass. Therefore, once β is so large that everyone has come together, they are all at the original center of mass. In this sense, there is no sensitive dependence on initial conditions. But the *path* to this outcome may depend sensitively on the initial parameters: for instance, the identity of the first pair to move may be changed, and this might dramatically change the path, although not the "final" outcome. We have not explored the precise nature of this sensitivity.

With quadratic moving costs, there is no (apparent) conserved quantity analogous to the center of mass in the linear case. Thus we suspect that changes in initial parameters that (say) change the identity of the first moving pair may affect the final location as well as the path.

7.5. Rationality

Without locations, perfect equilibrium, with its strong assumptions about rationality, and myopic-dynamic equilibrium coincide. This coincidence does not carry over with locations. For example, a group with perfect foresight might well want to move strategically, so as to pull others along in the near future. One move may lead to much faster network growth than another.[29]

8. Conclusions and Extensions

8.1. Summary of Findings

We have investigated some timing and target-selection issues in the theory of bandwagons driven by increasing network externalities. We explored the factors determining whether bandwagons form gradually or suddenly, and whether they form on the "right" target when groups differ in size, moving costs, and system quality. Finally, we explored the effects of horizontal differentiation ("locations").

Obviously, many questions remain. We pause to scratch the surface of two topics: strategic behavior in bandwagon industries, and the role of intellectual property rights and side payments in bandwagon industries.

8.2. Strategic Behavior in Bandwagons

Each group in our main model would like others to move to its system, and to do so as soon as possible. These desires suggest a number of strategic possibilities:

Entrenchment. Visibly raise one's own switching costs, as by investing or designing a system to be very different from others'. This makes it less likely that one will switch, and hence encourages others to switch to one's system instead.

Of course, this only works if one is a likely target. A very small group that entrenches itself only raises the moving costs that it will have to bear anyway, and harms everyone.

Growth. Attract more users to your group so as to solidify your position. It is especially good to attract influential users, whoever they are. How can this be done? Among attractive strategies may be penetration pricing, long-term contracts with buyers, and perhaps preannouncements if some buyers may thereby be persuaded to wait for your (forthcoming or soon-to-be-improved) system rather than adopting another now.

Proposition 6 has implications for the groups' incentives to grow in size, both absolutely and relative to one another. There is a discrete advantage to being the largest group, since that group bears no moving costs in equilibrium. In particular, given our equilibrium selection argument in the two-group case, group 1 has a distinct advantage over group 2, even if group 1 is only slightly larger than 2. This makes payoffs discontinuous at $s_1 = s_2$, which will create intense competition near there.

Quality Improvement. The incentive to improve the quality of one's system (if that is under one's control) has a strategic component in addition to the pure consumption incentive (make the system better for as long as one is using it). A quality improvement makes it more likely that others will switch to one's system, thus saving one's own moving costs.

Our results suggest that quality improvements are less effective than increases in size in this strategic respect, relative to the efficiency standard. That is, working for high quality is not as attractive as working for growth, because size is more powerful than quality in determining who becomes the bandwagon target.

Preemption. Pick a technology early on in the hope that others will follow, even though the early technology may be inferior.

8.3. Intellectual Property Rights and Side Payments

When one group moves to join another in our model, each confers a benefit on the other. One important set of questions concerns what happens if one or the other (or both) can charge for this benefit. As always, a perfect system of charging for benefits conferred will (tautologically) yield optimal behavior. More interesting is what will happen with realistic constraints on the system of payments. For example, one could consider various combinations of the constraints: i) one side can exclude the other but not vice versa; ii) subsidies but not charges are possible; iii) there is private information on groups' sizes and moving costs, and on systems' qualities.

Proposition 6 can also help us to think about side payments between the groups. While we have not formally modelled side payments, the result suggests that everyone might try to induce group two to move earlier: that would benefit both the largest group and all groups smaller than the second-largest. One can also ask whether group 1 can get away with charging others to join it (if it has the intellectual-property rights that entitle it to try). Given S, group 1 may be worse off the larger is s_2, for two reasons: a larger group 2 delays the avalanche *and* limits group 1's ability to charge a fee for access to its network.[30]

Appendix

Proof of Lemma

Suppose that j's equilibrium adoption date, T_j^* is earlier than i's adoption date, T_i^*. In equilibrium, the per-capita payoff to members of group j is

$$\int_0^{T_j^*} b_j e^{-rt} dt + \int_{T_j^*}^{\infty} [q(t) + \beta(t) v(S(t))] e^{-rt} dt - c_j e^{-rT_j^*}. \tag{15}$$

If, instead, group j were to adopt at date T_i^*, this payoff would become

$$\int_0^{T_j^*} b_j e^{-rt} dt + \int_{T_i^*}^{\infty} [q(t) + \beta(t) v(\hat{S}(t))] e^{-rt} dt - c_j e^{-rT_i^*}, \tag{16}$$

where $\hat{S}(t)$ is the alternative to $S(t)$ under j's deviation. Since j's delay cannot encourage others to adopt, and will have no effect on adoptions after t_i, we know that $\hat{S}(t) \leq S(t)$ for all t with equality for $t \geq T_i^*$.

To support such an equilibrium, the payoff in (15) must be greater than the payoff in (16). This condition implies that

$$\int_{T_j^*}^{T_i^*} [q(t) + \beta(t) v(S(t))] e^{-rt} dt > \int_{T_j^*}^{T_i^*} b_j e^{-rt} dt + c_j(e^{-rT_j^*} - e^{-rT_i^*}). \tag{17}$$

By similar reasoning, if i prefers to adopt at T_i^* rather than T_j^*, we must have

$$\int_{T_j^*}^{T_i^*} [q(t) + \beta(t) v(S(t))] e^{-rt} dt > \int_{T_j^*}^{T_i^*} b_i e^{-rt} dt + c_i(e^{-rT_j^*} - e^{-rT_i^*}). \tag{18}$$

To obtain (18) we have used the fact that earlier adoption by i can only accelerate adoption by others.

Using (17) and (18), we must have

$$\int_{T_j^*}^{T_i^*} b_i e^{-rt} dt + c_i(e^{-rT_j^*} - e^{-rT_i^*}) > \int_{T_j^*}^{T_i^*} b_j e^{-rt} dt + c_j(e^{-rT_j^*} - e^{-rT_i^*}).$$

Simplifying this expression gives

$$r(c_i - c_j) + (b_i - b_j) > 0,$$

which contradicts the assumption of the Lemma. ∎

Proof of Proposition 1. The Lemma tells us that the first movement must be by groups 1 through j for some j. T_j^* gives the first date at which groups 1 through j could coordinate a move to the new technology. By definition, the earliest such date, looking over all possible j's, is T_k^*. Under our assumption that groups can coordinate and move as soon as it is mutually beneficial to do so, movement must in fact occur at date T_k^*. Note also that there is no reason for groups 1 through k to delay: they find it worthwhile moving at T_k^* even if they are not followed in the future. Delaying would only reduce their flow payoff and (perhaps) delay others from joining, further reducing these agents' payoffs. ∎

Proof of Proposition 6. We prove this Proposition by induction on the number of groups, n. To do so, we decompose it into three related claims:

B(n): All movement is to the biggest group, group 1.
S(n): For some k, the first move is by $s_k, \cdots, s_n$ all moving together.
F(n): Let $\mathbf{s} = (s_1, \cdots, s_n)$ and let $\mathbf{s}' = (s_1 + \delta, s_2, \cdots, s_{k-1}, s_k - \delta, s_{k+1}, \cdots, s_n)$ for some $\delta > 0$. Then in the primed game, every group moves at or before the date at which they move in the unprimed game.

We prove these three claims by induction on n. First, for $n = 2$ they are either trivial or immediate consequences of our refinement assumption. Now, suppose that they hold for n and for all smaller numbers of groups; we must prove them for $n + 1$.

We will assume without further comment that the first-order condition actually characterizes the time of moving.

Proof of $B(n + 1)$. Could $B(n + 1)$ fail? Evidently, it would have to be the first movement that was to a nonconforming target, or else it would violate some $B(r)$ for $r \leq n$. So we ask: Can it be that the first movement in the $(n + 1)$-game is *not* to s_1?

We begin with the possibility that a single group, say group i, moves alone to somewhere other than group 1.

There are two subcases. First, if $i \neq 1$ and if either group i moves to an empty location i.e., creates an entirely new standard), or group i joins another group j such that $s_j + s_i < s_1$, then after the move group i is not part of the largest group.

To move, alone, to an empty location is ludicrous, since it leaves the game in exactly the same state as it had been but costs you moving costs. This strategy can therefore be ignored.

Suppose then that group i moves to join group $j \neq 1$, but does not thereby become part of the largest group, i.e., $s_j + s_i < s_1$. Then we have an n-game in which group i is not part of the largest group. By our inductive assumption, we know what will happen in that game. All movement will be to group 1, so group i gets a strictly lower flow than it would have done by joining group 1 originally, at least until the time when groups i and j move to group 1. But even then they get a lower flow, because comparing

$$(s_1 + s_i, s_2, \cdots, s_j, \cdots, s_n)$$

against

$$(s_1, s_2, \cdots, s_j + s_i, \cdots, s_n)$$

(where s_i is omitted in each case), we see that the former n-game has (weakly) faster movement than the latter, by property $F(n)$. Thus, even after group i bears the cost of moving again, it still does not get more than the flow benefit it would have been getting had it originally moved to 1 instead of to j. Hence, group i can do better by moving to join group 1 instead of to join group j.

What if by joining group $j \neq 1$, group i makes a new largest group, i.e., $s_j + s_i > s_1$? Then, just as if it had joined group 1, all subsequent movement is to group i. The comparison of its payoffs is then a matter of comparing the rate at which other groups join it. But comparing

$$(s_j + s_i, s_1, s_2, \cdots, s_{j-1}, s_{j+1}, \cdots, s_n)$$

versus

$$(s_1 + s_i, s_2, s_3, \cdots, s_j, s_{j+1}, \cdots, s_n),$$

we see that, by repeated application of $F(n)$, movement is (weakly) faster in the latter case. Hence again group i should join group 1 instead of group j.

This completes the proof that $B(n + 1)$ cannot be violated by a single-group first move. What about the possibility that several groups move at once, to a target $j \neq 1$? Intuitively, this cannot happen either, since the groups share a common interest exactly like group i in the analysis above. Indeed, one might argue that once they have decided all to move together, we can treat them as a single group, so if there are $m > 1$ of them we can just apply $B(n + 2 - m)$. We can't quite do that, because they don't get a chance to defect jointly (administrative coordination is by definition restricted to groups). However, if i is one of the groups meant to move to $j \neq 1$ at date T, then i can defect by moving to 1 at date $T - \varepsilon$ for small ε. By doing so, i ensures that it is part of the (new) largest group, so by $B(n)$ all subsequent movement will be to it: in particular, we know that all the other groups who were "meant" to move to j at T will now change their minds and move to 1 instead. Now the analysis is just as above.

This completes the proof of $B(n + 1)$. ∎

Proof of $S(n + 1)$. In the presence of $B(n + 1)$, $S(n + 1)$ is equivalent to the apparently weaker statement we shall call $\hat{S}(n + 1)$: "It cannot happen in equilibrium that a group of size s moves to group 1 at a date t, and then a strictly smaller group of size $s' < s$ moves to group 1 at a strictly later date $t' > t$."

Since we have established $B(n + 1)$, it is enough to prove $\hat{S}(n + 1)$. Let σ be the total size of the largest group including the group of size s and all who move at the same time, t. Then, since that group is willing to move at t, we must have:

$$\beta(t)(v(\sigma) - v(s)) \geq rc.$$

But this immediately implies that:
$$\beta(t)(v(\sigma + s') - v(s')) > rc,$$
so that the smaller group should move immediately at t (if indeed it should not have moved earlier). This contradiction establishes $\hat{S}(n + 1)$ and hence $S(n + 1)$. ∎

Proof of $F(n + 1)$. Consider the unprimed game, and suppose without loss of generality (by $B(n + 1)$ and $S(n + 1)$) that the first movement is by $s_j, s_{j+1}, \cdots, s_{n+1}$ all moving at t to 1.

By the condition that j is just willing to participate, we have:
$$\beta(t)(v(s_1 + s_j + \cdots + s_{n+1}) - v(s_j)) = rc.$$

If $j < k$ then there is no difference (between the primed and unprimed games) in the timing of the first move, and the state after the first move is the same, so everything is the same and $F(n + 1)$ holds (weakly).

If $j = k$ then $v(s_j)$ is smaller in the primed game, so j would sooner be willing too move (with $j + 1, \cdots, n + 1$) than in the unprimed game. Thus, either the first move in the primed game happens sooner than t, or (unlikely but possible) it happens exactly at t and involves more groups. In either case, the state after the first move is the same (between primed and unprimed), so again $F(n + 1)$ is confirmed.

Finally, if $j > k$ then $s_j + \cdots + s_{n+1}$ is the same between primed and unprimed, and since $s_1' > s_1$ the first move by groups $j, \cdots, n + 1$ can take place earlier; again, it is also possible that it could take place at t and involve more groups. But groups willing to move will never wait.

This proves $F(n + 1)$ and completes the proof of the Proposition. ∎

Proof of Proposition 8. Formally, we must show, given an arbitrary (x_i, s_i), that there does not exist β such that a triplet can come together but such that no pair can do so.

First, we *claim* that a necessary and sufficient condition for it to be an equilibrium for all three to coalesce is: each pair could coalesce were it to expect the third group then to come along and join them. Necessity is trivial, of course. To prove sufficiency, let R_i be the "furthest right" that group i would be prepared to go in order to join the other two, i.e.,
$$R_i = \max\{R \text{ such that } |R - x_i| \leq \beta(S - s_i)\} = x_i + \beta(S - s_i).$$
Similarly, let L_i be the furthest left that group i would go: $L_i = x_i - \beta(S - s_i)$. Let R^* be the minimum of the R_i, and L^* the maximum of the L_i. Then it is clear that all three groups can coalesce if and only if $R^* \geq L^*$. But a pair (i, j) can coalesce (on the presumption that the third group would join

them) if and only if $R_i \geq L_j$ and $R_j \geq L_i$. Since (trivially) $R_i \geq L_i$, this proves the claim.

Now for simplicity, call the groups' sizes (x, y, z) along with the locations left, middle, and right. Call the distance between the left and middle d and the distance between middle and right $1 - d$ (normalize the total distance to 1). Now the Proposition reduces to the statement that if β is large enough to enable the coalition of all three to form, i.e., (using our claim) if

$$\beta \geq \max\left\{\frac{d}{x+y+2z}, \frac{1}{x+2y+z}, \frac{1-d}{2x+y+z}\right\},$$

then also β is large enough that at least one of the three possible pairs would be willing to form, i.e.,

$$\beta \geq \min\left\{\frac{d}{x+y}, \frac{1}{x+z}, \frac{1-d}{y+z}\right\}.$$

That is, the Proposition claims that, for all $d \in (0, 1)$ and for all positive x, y, z,

$$\max\left\{\frac{d}{x+y+2z}, \frac{1}{x+2y+z}, \frac{1-d}{2x+y+z}\right\} \geq \min\left\{\frac{d}{x+y}, \frac{1}{x+z}, \frac{1-d}{y+z}\right\}.$$

We have "proved" this inequality this using simulation methods.[31] ■

Proof of Conjecture. By induction on N, it is enough to show that a pair will be able to form by the time the coalition of the whole can form. To find conditions for this to be true, note that by analogy with proof of the proposition for $N = 3$ that the grand coalition can form if and only if each pair would be willing to form if it believed it could drag all other groups along. The condition for this is that, for all i and j,

$$\beta \geq \frac{|x_i - x_j|}{2S - s_i - s_j}.$$

We want to show that this is inconsistent with β being so small that *no* pair (i, j) can form alone: i.e., for all i, j,

$$\beta < \frac{|x_i - x_j|}{s_i + s_j}.$$

Thus, what we must show is that

$$\min_{i,j}\left\{\frac{|x_i - x_j|}{s_i + s_j}\right\} \le \max_{i,j}\left\{\frac{|x_i - x_j|}{2S - s_i - s_j}\right\}, \tag{19}$$

or equivalently that

$$\max_{i,j}\left\{\frac{s_i + s_j}{|x_i - x_j|}\right\} \ge \min_{i,j}\left\{\frac{2S - (s_i - s_j)}{|x_i - x_j|}\right\}. \tag{20}$$

Several attempts (by us and by helpful others) to construct counterexamples to (19) have failed, so we tend to believe it. (Notice that for a particular i,j the left hand side may easily exceed the right hand side, so the min and the max are important.)

This result would not hold if the d_{ij} were not constrained to be given by $|x_i - x_j|$'s for real x_i. For instance, in $N - 1$ dimensions, it is easy to ensure that $|x_i - x_j| = 1$ for all $i \ne j$, and then counterexamples are readily constructed. Thus we view the "result" as suggestive rather than in any way general. ∎

Notes

1. "Incompatibility" in the information sector means a variety of things, notably that documents or data created using one system cannot easily be transferred to another. Of course, inexpensive translators or converters can sometimes obviate the need to adopt compatible technologies. See Farrell and Saloner (1992) for an economic analysis of converters.

2. See e.g., *Business Week*, August 20, 1990, "Bolting from the Cellular Herd," and April 27, 1992, "A Double Standard for Cellular Phones."

3. See, for example, Arthur (1989), Farrell and Saloner (1985 and 1986), and Katz and Shapiro (1985 and 1986a).

4. See Katz and Shapiro (1985) for the argument that incompatibility may be profitable for some vendors; see Klemperer (1987) and Farrell and Shapiro (1988) for models suggesting that raising switching costs for consumers may be profitable for vendors.

5. A "large" group could be one with many members, or one with members who intensively use the product or service under study. For ease, we shall speak in terms of size rather than intensity of use, but either interpretation fits our analysis.

6. Although our analysis includes the possibility that network externalities are important for the old technology as well as the new, we do not include stranding externalities—diminished benefits to one group as others cease using its technology—in our analysis. Stranding externalities are absent if each group initially uses a separate older technology, or if the network benefits depend upon the maximum number of users who have ever used a technology (rather than the number currently doing so).

7. For simplicity, we rule out side payments between agents. In the conclusion we discuss some of the issues raised if such side payments are feasible.

8. In future work we plan to explore bandwagon behavior when quality is frozen as of the date of first adoption of the new technology.

9. This implies that the strategy spaces are large, even though (as remarked above) each agent is "merely" choosing a single adoption date.

10. The additional complexity is not great if moving costs vary with time—the formulation $c_i(t)$. Essentially, the agent must still calculate the extra (present value) adoption costs associated with adopting earlier. These flow costs are no longer rc_i, but must be modified to $rc_i(t) - c'_i(t)$.

11. See Cabral (1989) and Agliardi (1992), for instance.

12. We obtain this result because we have omitted any "stranding externality." More generally, if not all users of an old technology move at the same time then stranding may be important, and the direction of the bias is ambiguous.

13. This observation suggests to us that further analysis including the possibility of side payments is important.

14. Potentially, a group could move more than once. This raises questions about its second-time moving cost. In our model, groups will not in fact move repeatedly.

15. In this specification, a user's benefits do not depend on the particular system she picks, but only on the date and the size of the network to which she belongs. Below we will generalize the model to permit systems to differ in quality.

16. For consistency, suppose that benefits will be enjoyed over time, just as in the dynamic model, but there is only one date at which movement can occur.

17. When the technologies are of sufficiently unequal quality, the waiting game structure may not apply because the group with the inferior technology might rather switch than be switched to. In that case, the game is trivial.

18. By this term we will mean, throughout, the time that group 2 would optimally move if it believed that no other group would move in the future. We do not mean, for instance, the first time at which group 2 would prefer to move rather than to swear never to move.

19. There may exist a mixed-strategy equilibrium. In any such equilibrium, group 2 might move with probability less than 1 at T_2^* but there will be no chance of movement at dates strictly between T_2^* and T_1^*. At T_1^*, group 1 might move (again, with probability less than 1); thereafter, there would be complex randomizations. We do not pursue this here, since we focus on a pure-strategy equilibrium that always exists.

20. We shall, however, later discuss the possibility that influential agents can make the other equilibrium focal.

21. That is, group 2 can "signal" by delaying after T_2^* (indeed, until T_1^*) that it does not mean or expect to play the first equilibrium; some theorists would argue that this makes the second equilibrium (which, when it exists, is better for group 2) more plausible than the first. We find this unconvincing in this context.

22. In high-definition television, the FCC has announced a policy of forcing NTSC viewers to switch to HDTV by 2004. In this case the welfare costs of incompatibility are primarily duplication of TV signals, which uses more electromagnetic spectrum than would otherwise be needed.

23. This latter effect is like the role played by celebrities in advertising. The notion of trend setters or opinion leaders is well known in marketing.

24. We do not explore the timing of the new user's choice.

25. This discussion is similar to that in Farrell and Saloner (1992, page 24).

26. This less popular technology must, in this case, have higher moving costs, since otherwise T_1^* and T_2^* could not be equal.

27. See Farrell and Saloner (1986b, 1992) for static analysis of compatibility versus variety in a horizontally differentiated model.

28. A more general model would include partial compatibility, where the network externalities between two groups would be greater, the closer are those two groups.

29. We have postulated that each group can move at most once. With multiple moves, another divergence appears between myopic dynamics and perfect foresight: it is unwise for a group to ignore the fact that it will move again in the future. The group might be wise to avoid a move that was predicated on amortizing the moving costs indefinitely.

30. Of course, a competitive group 2 might also be charging and/or subsidizing, so...

31. We normalized $y = 1$, and then did two searches over x, and z. One had these sizes run from 0 to 100 by 1, the other ran them from 0 to 10 by 0.1. In each case we let d vary by 0.01 on the unit interval. In some extreme cases, the two expressions were equal, but generally the maximum was strictly bigger than the minimum.

References

Agliardi, Elettra. 1992. "Discontinuous Adoption Paths with Dynamic Scale Economies," Mimeo, Cambridge University.

Arthur, Brian. 1989. "Competing Technologies, Increasing Returns, and Lock-In by Historical Small Events," *The Economic Journal* 99: 116–131.

Cabral, Luis. 1989. "On the Adoption of Innovations with Network Externalities," Unpublished manuscript: Lisbon University.

Farrell, Joseph, and Garth Saloner. 1985. "Standardization, Compatibility, and Innovation," *Rand Journal of Economics* 16: 70–83.

Farrell, Joseph, and Garth Saloner. 1986. "Installed Base and Compatibility," *American Economic Review* 76: 940–955.

Farrell, Joseph, and Garth Saloner. 1992. "Converters, Compatibility, and the Control of Interfaces," *Journal of Industrial Economics*, 9–36.

Farrell, Joseph, and Carl Shapiro. 1988. "Dynamic Competition with Switching Costs," *Rand Journal of Economics* 19: 123–137.

Katz, Michael L., and Carl Shapiro. 1985. "Network Externalities, Competition and Compatibility," *American Economic Review* 75: 424–440.

Katz, Michael L., and Carl Shapiro. 1986a. "Technology Adoption in the Presence of Network Externalities," *Journal of Political Economy* 94: 822–841.

Katz, Michael L., and Carl Shapiro. 1986b. "Product Compatibility Choice in a Market with Technological Progress," *Oxford Economic Papers, Special Issue on the New Industrial Economics* 38: 146–165.

Klemperer, Paul. 1987. "Markets with Consumer Switching Costs," *Quarterly Journal of Economics* 102: 375–394.

9 DYNAMIC TARIFF GAMES WITH IMPERFECT OBSERVABILITY

Andreas Blume and Raymond G. Riezman

1. Introduction

As observable tariffs come down, hidden protection becomes increasingly important. There is strong intuition that imperfect observability of actions e.g., in the form of hidden protection, may impede the attainment of efficiency in trade between countries even if the trading relationship is not of short duration. It is also well known that imperfect observability per se need not lead to inefficiency in repeated relationships. This suggests the following questions. What makes the tariff problem distinct from other repeated relationships? Does this distinction make it harder to achieve efficiency? How does one model this distinction formally? Are there policies to address these problems? Are these policies simple?

We want to explore how one can bring the tools and results from repeated game theory to bear on these issues. We consider dynamic tariff models in which countries cannot perfectly monitor the policies of their trading partners. Instead they condition their strategies on some publicly observable variable.

In section 2 we argue that the Prisoners' Dilemma game captures essential aspects of the tariff problem. In section 3 we will show in a simple example of a repeated PD game with discounting and imperfect observability, how certain types of strategies which achieve efficient outcomes without imperfect observability fail to do so when information about actions is only indirectly available through their effect on the distribution of a publicly observable variable. Of course, the inefficiency may disappear once one admits more general classes of strategies. In fact, in section 4, we describe a Folk Theorem (FT) for repeated games with imperfect observability which was obtained by Fudenberg, Levine and Maskin (1991). Subject to a (fairly weak) condition that defections can be statistically identified, they show that any individually rational payoff can be supported in equilibrium for a sufficiently high discount factor. The special structure of the tariff problem suggests the possibility that any distribution of the observable which can be induced by a domestic tariff policy, can be replicated by a foreign tariff policy and vice versa. This seems to contradict the above mentioned statistical identifiability condition which Fudenberg, Levine and Maskin show to be sufficient for a Folk Theorem in these models. Riezman (1991) in a specific version of the tariff problem, with strategies only conditioning on the history of the terms of trade, has raised the possibility that unlike for example in the Green and Porter (1984) setting, it may be impossible to sustain any cooperation with symmetric trigger strategies. In section 5 we will show that the Fudenberg, Levine and Maskin Folk Theorem applies to a general class of reduced form tariff games with imperfect observability. Thus, in these games imperfect observability does not restrict efficiency in the limit as the discount factor becomes large, as long as no strategies are excluded. We show that the Folk Theorem does hold under fairly general conditions if action spaces in the stage game are restricted to be finite. However, the FT is silent about how complex strategies need to be in order to achieve close to efficient outcomes. The complexity issue is explored in sections 6 and 7. To facilitate this task we assume in both sections that players can condition their actions on a very simple correlation device in addition to the publicly observable history. In section 6 we ask what is achievable with symmetric strategies. We compute a bound on time-average payoffs from symmetric equilibrium strategies which is independent of the discount factor and below the efficient symmetric stage game payoffs. Thus, not even in the limit, as the discount factor converges to one, is efficiency achievable with symmetric strategies. This suggests that occasional trade wars where punishment is applied indiscriminately are not conducive to the achievement of efficient outcomes. In section 7 we examine a class of asymmetric strategies which can be implemented with three state variables. The three states are a cooperative state and one punishment state for each player. This setting is clearly restrictive, in that it does not allow for, say, delayed punishments or any other strategy which involves counting. However many of the effects achievable with such strategies can be simulated via the above mentioned correlation device. For example, while delaying punishment is one way to exercise control over the severity of punishment, reducing the probability of entering a punishment phase is another. We believe that the employment of a correlation device is an effective way of exploring, certainly not all, but a large range of strategic possibilities. In this setting we show that again there is a bound on the symmetric payoffs obtainable in the class of asymmetric strategies considered. This bound does not depend on the discount factor and is below the efficient symmetric stage game payoffs.

Our analysis is somewhat in the spirit of Radner, Myerson and Maskin (1986). In both cases inefficiencies arise because punishments are invoked with positive probability even if all players conform with the equilibrium profile and because punishment payoffs must be below the efficient frontier. In their case the latter fact was true because the structure of the game made it impossible even statistically to distinguish who might have defected. In our setting the problem arises because of restrictions on the strategy space, e.g. in the case of symmetric strategies it is evident that any punishment affects both players and if invoked with positive probability on the equilibrium path reduces both players expected payoff. Radner, Myerson and Maskin do not restrict strategies while our observable is more informative about who cheated than theirs.

Abreu, Pearce and Stacchetti (1986, 1990) address the question of the complexity of constrained efficient strategies in repeated games with discounting and imperfect observability. They show that under their assumptions after every history of the observable the continuation value of an efficient equilibrium strategy is an extreme point of the equilibrium value set. In the case of symmetric strategies, this implies that only two action profiles will ever be used in equilibrium.

Our approach to the issue of complexity asks how well one can approximate constrained efficient payoffs with simple policy rules. One can think of such rules in terms of strategies which can be implemented via finite automata (e.g., Abreu and Rubinstein, 1988). To this we add a simple publicly observable random variable. The larger the number of states an automaton is allowed to have, the more strategies we can construct. We show that in an example of a prisoners' dilemma game that the payoffs from a large class of strategies which use no more than three states is bounded away from the efficient frontier. We conjecture but cannot prove that this result holds for all finite automata with a fixed finite number of states.

The complexity of strategies needed to obtain efficiency in repeated games varies with the assumptions on repeated game payoffs, observability, etc. In the undiscounted case it may be necessary to use strategies with punishments which "fit the crime," i.e., depend in complicated ways not only on who defected but on the specific defection and the history prior to the defection (compare Rubinstein, 1979). The discounted case with perfect observability is easier in the regard. Abreu (1988) shows that in these games simple strategy profiles suffice to support all payoffs from subgame-perfect equilibria. Fudenberg and Maskin's (1986) proof of the folk theorem for repeated games with discounting and perfect monitoring is constructive; simple stick-and-carrot strategies suffice for the argument. This is not the case with imperfect observability. Fudenberg. Levine and Maskin (1991) characterize the equilibrium payoff set directly without constructing strategies explicitly (see also Matsushima, 1989). Our paper suggests that explicitly constructing strategies which approximate efficient payoffs in these games may be quite difficult. If this conjecture is correct one should be cautious about invoking "the folk theorem" in applications.

2. The Trade Model

We consider a simple two country-two good pure exchange model of international trade. There are two goods, X and Y and two countries home (h) and foreign (f). Each country has identical consumers with the utility function

$$U^i = X^i Y^i \ i = h, f.$$

Define the world endowment of each commodity to be one unit, then country h is endowed with γ units of X and country f has $1 - \gamma$; country f has μ units of Y and country h has $1 - \mu$ units, where $\gamma, \mu > 1/2$. In equilibrium, h exports X and imports Y. Country h charges tariffs of $S - 1$ on Y and f charges $T - 1$ on imports of X. Solving for the offer curves we get for country h

$$\frac{\gamma}{X} = \frac{S(1-\mu)}{Y} + S + 1$$

The foreign country's offer curve is given by

$$\frac{\mu}{Y} = \frac{T(1-\gamma)}{X} + T + 1$$

Solving the offer curves one gets expressions for equilibrium consumption and hence utility. For country h

$$U^h = \frac{(\gamma + (1-\mu)T)^2}{(1+(1-\mu)T + \mu S^{-1})(T + (1-\gamma)ST + \gamma)}$$

$$U^f = \frac{(\mu + (1-\gamma)S)^2}{(1+(1-\gamma)S + \gamma T^{-1})(S + (1-\mu)ST + \mu)}$$

Then each country solves an optimal tariff problem and chooses the tariff that maximizes the utility of a representative consumer subject to a balance of trade constraint. The first order conditions for this problem yields the tariff reaction functions for the two countries. For country h

$$\frac{\mu}{S^2(1+(1-\mu)T+\mu S^{-1})} = \frac{1-\gamma}{1+(1-\gamma)S+\gamma T^{-1}}$$

Country f's reaction function is defined by

$$\frac{\gamma}{T^2(1+(1-\gamma)S+\gamma T^{-1})} = \frac{1-\mu}{1+(1-\mu)T+\mu S^{-1}}$$

It is now possible to compute tariffs and utilities for different strategy combinations. If country f practices free trade under all circumstances then the optimal tariff for h is given by the usual optimal tariff formula which for this case is

$$S_0 = \sqrt{\frac{\mu(1+\gamma)}{(\gamma-1)(\mu-2)}}$$

The optimal tariff for f is similarly defined is

$$T_0 = \sqrt{\frac{\gamma(1+\mu)}{(\gamma-2)(\mu-1)}}$$

If both countries charge tariffs the Nash equilibrium tariffs are

$$S_N = \sqrt{\frac{\mu}{1-\gamma}}$$

$$T_N = \sqrt{\frac{\gamma}{1-\mu}}.$$

One can substitute these values for tariffs into the utility functions above to get expressions for utility in terms of the initial endowments. In order to transform this game into one with a finite strategy space, suppose that both countries' choices are restricted to free trade or to the tariff which would be a best response to free trade. Then, by varying the endowments, different payoff matrices are generated. There is another Nash equilibrium in the original game which we ignore. If tariffs are set high enough to eliminate trade, autarky results. Autarky is a Nash equilibrium.

There are two possible payoff structures. If both countries do worse at the Nash equilibrium than at free trade then the game has a prisoner's dilemma structure. We will concentrate on this case. However, it is possible that one country does better at Nash equilibrium than it does at free trade. This will be true if one country is large relative to the other (see Kennan and Riezman [1988]).

We next look at a typical payoff matrix for the prisoner's dilemma game. If $\gamma = .7$ and $\mu = .6$ then we get the following payoff matrix

	FT	T
FT	302, 202	290, 208
T	313, 189	299, 194

Notice that in this game, both countries have higher utility if they charge tariffs than if they do not, independent of the other country's strategy. We next consider this game played repeatedly.

3. Inefficiency with Restricted Strategy Spaces: An Example

Consider the prisoners' dilemma game.

	L	R
U	5, 5	0, 7
D	7, 0	1, 1

Player one chooses rows and player two chooses columns. Each cell in the payoff matrix specifies first the row player's payoff and then the column players payoff. With the following interpretation this becomes an example of a tariff game. Let U stand for a low and D for a high tariff policy. Take for a example the case where U corresponds to free trade and D to the myopic best response to a free trade policy by the other country.

Let this game be repeated infinitely often and assume that after each round players can observe which out come was realized in that round.

Strategies in the infinitely repeated game are functions from observable histories into each player's set of actions. Any pair of strategies induces a path of actions and a path of associated stage game payoffs in a natural way. Each player's payoff from a pair of repeated game strategies is equal to the expected discounted sum of the sequence of stage game payoffs induced by those strategies. Assume that both players have a common discount factor δ. That is, if a player's stage game payoffs are π_t in period t, then her repeated game payoffs are given by

$$(1-\delta)\sum_{t=0}^{\infty}\delta^t \pi_t.$$

The factor $1 - \delta$ enters because we want to compute an average; e.g., a constant stream of payoffs π will yield a repeated game payoff of π as well.

It is a simple fact that for high enough discount factors the efficient symmetric outcome can be supported with the threat of "(D, R) forever" or with milder strategies which return to cooperation after a punishment period of finite length, so called trigger strategies. Now modify this game as follows. Let $P(x|y)$ denote the probability that the effective policy is x when the intended policy is y and assume $P(x|x) = \alpha$, $x = U, D, L, R$. Let $\alpha > 1/2$. For the tariff problem this simple form of imperfect observability might be justified as arising from the difference between political intent and the actual implementation of a policy. For example, a low tariff could be agreed upon by a country's negotiators and passed by the legislature. However, this policy reform could be undone by a government agency that finds evidence of dumping, thereby allowing the tariff to be reinstituted. There could also be instances in which government agencies adopt policies (health and safety guidelines for example) which, while not trade policies per se, have a definite impact on trade. In practice, a large number of policies and agencies influence variables which ultimately have an effect on trade.

For a high enough discount factor δ, the trigger strategy which after any observation of (D, L) or (U, R) outside a punishment phase punishes for one period with (D, R) can be shown to sustain some cooperation. The time average equilibrium payoff from that strategy equals

$$v_c^{sym} = \frac{\alpha^2(1-\delta)5 + (1-\alpha)^2(1-\delta) + \alpha(1-\alpha)[(1-\delta)7 + 2\delta(1-\delta)B]}{1 - \delta\alpha^2 - \delta(1-\alpha)^2 - 2\delta^2\alpha(1-\alpha)}$$

where

$$B = \alpha^2 + 5(1-\alpha)^2 + 7\alpha(1-\alpha)$$

The highest symmetric time average payoff obtainable is given by
$$A = 5\alpha^2 + (1 - \alpha)^2 + 7\alpha(1 - \alpha)$$
In the limit, as the discount factor approaches unity, the time average payoff from the trigger strategy converges to
$$\lim_{\delta \to 1} v_c^{sym} = \frac{1}{1 + 2\alpha(1 - \alpha)} A + \frac{2\alpha(1 - \alpha)}{1 + 2\alpha(1 - \alpha)} B.$$
It is easy to simplify this expression further. However, this way of writing it shows that the limiting payoff is a strictly convex combination of A and B. Since $B < A$, it is strictly less than A for $1/2 < \alpha < 1$. Thus, even in the limit, as $\delta \to 1$, time average payoffs from the mildest trigger strategy are bounded away from the efficient frontier.

This observation however is a little misleading because one can do better by admitting asymmetric punishments. Consider the following strategy for the row player: Let history be summarized by three states C, Q_1, Q_2, a cooperative state, a punishment state for player 1 and a punishment state for player 2. In state C the strategy prescribes U, in state Q_1 it prescribes U and in state Q_2 it prescribes D. The state transition rule is as follows. In C return to C after (U, L) and (D, R), go to Q_1 after (D, L) and to Q_2 after (U, R). In Q_1 return to Q_1 after (D, R); otherwise return to C. In Q_2 return to Q_2 after (D, R); otherwise return to C. Suppose that the column player uses the same strategy except that in Q_1 the column player uses R and in C and in Q_2 she uses L. Let
$$B_1 = 7(1 - \alpha)^2 + 6\alpha(1 - \alpha)$$
$$B_2 = 7\alpha^2 + 6\alpha(1 - \alpha)$$
Like before, let v_c^{asym} denote the time average payoffs from the above strategy combination. One can show that
$$\lim_{\delta \to 1} v_c^{asym} = \frac{1 - \alpha(1 - \alpha)}{1 + \alpha(1 - \alpha)} A + \frac{\alpha(1 - \alpha)}{1 + \alpha(1 - \alpha)} B_1 + \frac{\alpha(1 - \alpha)}{1 + \alpha(1 - \alpha)} B_2.$$
Observe that
$$B < \frac{B_1 + B_2}{2}.$$
Also for α close to one the weight on B is close to the weight on $\frac{B_1 + B_2}{2}$. For sufficiently high α,
$$\lim_{\delta \to 1} v_c^{asym} > \lim_{\delta \to 1} v_c^{sym}.$$

This shows that asymmetric punishments lead to higher payoffs. However, the limiting payoff, as $\delta \to 1$, from the asymmetric profile is still less than the efficient symmetric stage game payoff A because $\frac{B_1 + B_2}{2} < A$. Therefore, while asymmetric punishments yield higher payoffs than symmetric punishments they still are lower than the efficient stage game payoff.

Here, as throughout the paper the efficiency standard is constrained efficiency. We are interested in strategies which approximate payoffs on the efficiency frontier of the set of payoffs from perfect public equilibria. It is hard to say much about the latter set in general. Therefore we consider limits as the discount factor approaches 1 and invoke a folk theorem.

4. The Folk Theorem in Repeated Games with Imperfect Observability

The following description of repeated games with imperfect observability and the statement of the sufficient conditions for a folk theorem to hold in these games is borrowed from Fudenberg, Levine and Maskin (1991) A_i denotes player i's set of actions in the stage game and m_i is the cardinality of this set. Let α denote a mixed strategy profile and let α_{-i} denote those components of α which exclude player i. In the theory developed by Fudenberg, Levine and Maskin a player's payoff may depend on a publicly observable variable as well as a private observable. A public strategy is a strategy which conditions only on past realization of the public random variable. Fudenberg, Levine and Maskin consider *perfect public equilibria* (PPE). These are profiles of public strategies which form Nash equilibria after all public histories; notice that this requires stability also against defections to non-public strategies. Denote the finite set of public observables by Y and the finite set of private observables by Z. Let $\pi(\cdot, \cdot | \alpha)$ denote the joint probability distribution which a mixed strategy profile α induces on the set $Y \times Z$. The marginal distribution of the public observable is written as $\pi(\cdot | \alpha)$. Let $\overline{m}$ be the cardinality of Y. Then $\pi(\cdot | \cdot, \alpha_{-i})$ is the matrix of distributions generated if player i's strategy is allowed to vary over the set of her actions a_i and the other players' mixed strategies are fixed at α_{-i}. Players are assumed to discount stage game payoffs with a common discount factor δ.

Definition 1. *The profile α has individual full rank for player i if the m_i vectors $(\pi(\cdot | \cdot, \alpha_{-i}))_{a_i \in A_i}$ linearly independent. It has individual full rank if it has the above property for every player i.*

Definition 2. *Given a pair of players $i \neq j$, α satisfies pairwise full rank for (i, j) if the matrix*

$$\begin{pmatrix} \pi(\cdot|\cdot, \alpha_{-i}) \\ \pi(\cdot|\cdot, \alpha_{-j}) \end{pmatrix}$$

has rank $m_i + m_j - 1$.

Individual full rank guarantees that defections are detectable in a statistical sense. Similarly, pairwise full rank ensures that in a statistical sense defections of any two players can be distinguished from each other; notice that $m_i + m_j - 1$ is the maximal possible rank of the matrix in definition 2.

Two conditions are sufficient for the FT. These are pairwise full rank at *one* profile and individual full rank at all pure profiles.

Condition P. *A game satisfies the pairwise full rank condition if, for all pairs (i, j), $i \neq j$, these exists a profile α that satisfies pairwise full rank for that pair.*

Condition I. *A game satisfies the individual full rank condition if any pure action profile has individual full rank.*

Notice that condition P in particular is a weak condition because pairwise full rank needs to hold at only one mixed profile and the profile can be chosen conditional on the pair of players. Let V^* denote the set of feasible individually rational payoffs, the set of payoffs below which no player can be held by the others. Lee $E(\delta)$ stand for the set of equilibrium payoffs from perfect public equilibria in the infinitely repeated game when stage game payoffs are discounted with a factor δ.

Fudenberg, Levine and Maskin [FLM] prove the following result.

Theorem [FLM]. *Suppose that conditions P and I are satisfied. Let $\hat{V}$ be a smooth subset of the interior of V^*. Then there exists a $\underline{\delta}$ such that for all $\delta > \underline{\delta}, \hat{V} \subseteq E(\delta)$.*[1]

5. The Folk Theorem in the Reduced Form Trade Model

In this section we consider a bilateral tariff game in reduced form a la Dixit (1987) and construct a worst case scenario for an FT. If we can nevertheless show that an FT holds, then there is no reason why efficiency should be ruled out in principle and we can proceed with examining the effects of limiting strategic complexity.

The reason for constructing such a worst case scenario is that as Riezman (1991) pointed out it may be impossible to sustain any cooperation with symmetric strategies in a tariff game if countries condition only on the terms of trade. In order to see how serious this problem is, one must disentangle the specifics of the tariff game problem from the strategic complexity issue. An important characteristic of Riezman's setup was that any distribution one of the countries could induce by raising its tariff, the other could induce as well by lowering its tariff. This seems to contradict Fudenberg, Levine and Maskin's pairwise full rank condition. We want to check whether under similar conditions (with finitely many actions) it is possible to attain efficiency in the limit with more general strategies.

By a worst case scenario we mean the following: First, we restrict the set of observable outcomes, by assuming that quantities are only privately observable. This is best thought of as a restriction on strategies and might be motivated by arguing that the political process necessitates that only relatively simple strategies be considered. Leaving only the terms of trade as an observable is perhaps not very realistic but at least conceptually poses a special problem: With enough symmetry any distribution over the terms of trade which one country can induce with one of its tariff policies the other could have induced as well. Thus, if there is a finite model with such symmetry, the pairwise full rank condition must fail. The second component of our worst case scenario are restrictions on the distribution of the terms of trade in the spirit of the above mentioned symmetry. In the end it will turn out that finiteness (which is best thought of as boundedness) imposes restrictions on how much symmetry one can demand and thus the FT continues to hold. Boundedness of the terms of trade is not unreasonable, considering that extremely unfavorable terms of trade would correspond to one country subsidizing the other which seems politically infeasible.

The following argument establishes the Folk Theorem for a bilateral tariff game in reduced form in which the terms of trade p are directly expressed as functions of the domestic tariffs, $t \in T$, and foreign tariffs $t^* \in T$.[2] It is without loss of generality to identify the finite set T with a set of the form $\{1, 2, \cdots, |T|\}$, and similarly for T^*. With imperfect observability, the t.o.t depend in addition on domestic shocks $s \in S$ and foreign shocks $s^* \in S^*$. We will assume $S \supset T$ and $S^* \supset T^*$ as part of a worst case scenario for the FT. Thus the terms of trade for a given set of shocks and tariffs (s, s^*, t, t^*) are given by $p(s, s^*, t, t^*)$. Assume that

$$p(s, s^*, t+1, t^*) > p(s, s^*, t, t^*) \; \forall (s, s^*, t, t^*)$$

and

$$p(s, s^*, t, t^*+1) < p(s, s^*, t, t^*) \; \forall (s, s^*, t, t^*)$$

196 PROBLEMS OF COORDINATION IN ECONOMIC ACTIVITY

We also want to capture the idea that 1. tariffs may be masked by shocks and that 2. for a given pair of shocks from the status quo each country can on its own generate a given t.o.t unless this violates the bounds on tariffs implicit in the finiteness of T and T^*. This is expressed in the following conditions

$$p(\theta, s^*, \tau, t^*) = p(\tau, s^*, \theta, t^*)$$
$$p(s, \theta^*, t, \tau^*) = p(s, \tau^*, t, \theta^*)$$

and

$$p(s, s^*, t+j, t^*) = p(s, s^*, t, t^*-j) \quad \forall j : t+j \in T \text{ and } t^*-j \in T^*$$

Aside from these conditions we will assume that all t.o.t's are different. Let $\underline{t} = \min T$ and $\bar{t} = \max T$ with similar conventions on shocks and the corresponding foreign country's variables. Then

$$p(s, s^*, \underline{t}+k, \underline{t}^*+l) > p(\underline{s}, \bar{s}^*, \underline{t}+k-l-1, \underline{t}^*) =$$
$$p(\underline{s}, \underline{t}^*, \underline{t}+k-l-1, \bar{s}^*) = p(\underline{s}, \underline{t}^*, \underline{t}, \bar{s}^*-k+l+1) =$$
$$p(\underline{s}, \bar{s}^*-k+l+1, \underline{t}, \underline{t}^*)$$

Similarly,

$$p(s, s^*, \underline{t}+l, \underline{t}^*+k) < p(\bar{s}, \underline{s}^*, \underline{t}, \underline{t}^*+k-l-1) =$$
$$p(\underline{t}, \underline{s}^*, \bar{s}, \underline{t}^*+k-l-1) = p(\underline{t}, \underline{s}^*, \bar{s}-k+l+1, \underline{t}^*) =$$
$$p(\bar{s}-k+l+1, \underline{s}^*, \underline{t}, \underline{t}^*)$$

These inequalities identify the highest and lowest prices; e.g., at free trade, the lowest price is $p(\underline{s}, \underline{s}^*, \underline{t}, \underline{t}^*)$. Any lower price has probability 0. Tariffs other than the free trade tariffs $(k, l \neq 0)$ shift these upper and lower bounds. Then assume that all prices which are not ruled out by these restrictions occur with positive probability (without any other restrictions on these probabilities) and form the matrix.

$$\{Prob\{p(r, r^*, \underline{t}, \underline{t}^*) \mid (\underline{t}, \underline{t}^*)\}\}_{((t,t^*)=(\bar{t},\underline{t}^*)\ldots(\underline{t},\underline{t}^*)\ldots(\bar{t},\underline{t}^*)),((r,r^*) = (\underline{s},\underline{s}^*+\bar{t}^*)\ldots(\underline{s},\underline{s}^*)\ldots(\bar{s}+\bar{t},\underline{s}^*))}$$

If we let $* * *$ denote a $1 \times (|S| + |S^*|)$ vector of positive numbers which sum to one, then the inequalities (3.a) and (3.b) imply that this matrix is of the form

	$(\underline{s}, \bar{s}^* + \bar{t}^*)$	.	.	.	.	.	.	.	.				$(\bar{s} + \bar{t}, \underline{s}^*)$
$\bar{t}, \underline{t}^*$	0	.	.	.	.	.	.	.	0	*	*	*
	0	.	.	.	.	.	.	.	0	*	*	*	0
.	.	.	.	.	.	.	.	.	.	.	.	.	.
	0	.	.	.	0	*	*	*	0	.	.	0
$\underline{t}, \underline{t}^*$	0	.	.	.	0	*	*	*	0	.	.	0

```
   .          0       .   .  0  *  *  *  0   .   .   .   .        0
   .          .       .   .  .  .  .  .  .   .   .   .   .        .
   .          0            *  *  *  0  .   .   .   .   .          0
t, t̄*         *            *  *  0  .   .   .   .   .   .          0
```

The previous matrix has rank equal to $|T| + |T^*| - 1$. This shows that our bilateral tariff game has pairwise full rank at the free trade profile ($\underline{t}$, $\bar{t}^*$). In-dividual full rank at all pure strategy profiles is clearly satisfied, because any unilateral change in strategy shifts the support of the distribution. Following Fudenberg, Levine and Maskin this yields the following result.

Proposition. *The Folk Theorem holds in the reduced form bilateral tariff game with imperfect observability.*

6. Optimal Symmetric Strategies

Having shown that in principle efficiency is attainable in the tariff game, we will now continue to analyze restrictions on the complexity of strategies and their effect on efficiency. Our goal is to try get a feel for how complex strategies have to be in order to attain an efficient outcome. For this purpose we will give up the general framework and instead work with the PD game whose relation to the tariff problem we have established earlier. It is also easy to verify that the imperfect observability version of the PD game introduced in section 3 satisfies the conditions for the Fudenberg, Levine and Maskin folk theorem. In this section we consider a general class of prisoners' dilemma games with imperfect observability and compute an upper bound on payoffs from symmetric strategies. We show that the upper bound is independent of the discount factor and below the efficient symmetric payoff. Thus, even as the discount factor approaches unity, payoffs from symmetric strategies are bounded away from the efficient frontier. The upper bound of the symmetric payoff which we compute can be attained as an equilibrium payoff in the game if players have a simple publicly observable randomizing device available. The strategy which we construct involves punishing with some probability if certain outcomes are observed. Interestingly, in any strategy which attains the best symmetric payoff, punishment occurs with positive probability only in the event "both defect," which is in stark contrast to the perfect observability case where punishment needs to be initiated only in the event of a single defection. The model is essentially the same as section 3 except that we consider a general prisoner's dilemma game.

In the following we assume that players can publicly observe the realizations of a random variable which are uniformly distributed on the unit interval. Consider the following class of prisoners' dilemma games and retain the structure from section 3.

$$\begin{array}{c|cc} & L & R \\ \hline U & e,e & f,g \\ D & g,f & h,h \end{array}$$

where $g > e$, $e > h$, $h > f$, $2e > g + f$. Since payoffs are bounded there exists an upper bound on payoffs from symmetric perfect public equilibria. In the appendix we show that one can construct a PPE strategy which attains the upper bound. This maximum value v_c from symmetric equilibrium strategies can be decomposed into the first period payoff plus the continuation value, for any strategy σ supporting v_c. Clearly, if there is any cooperation under such a strategy, then the first period payoff must be the expected payoff from cooperation because otherwise we could just start the game in a different "subgame." We will assume that the discount factor is sufficiently close to one to ensure that some cooperation is feasible and that α is large in a sense that is made precise below. Hence,

$$v_c = (1 - \delta)[e\alpha^2 + f\alpha(1 - \alpha) + g\alpha(1 - \alpha) + h(1 - \alpha)^2] \\ + \delta\alpha^2 v_e + \delta\alpha(1 - \alpha)v_f + \delta\alpha(1 - \alpha)v_g + \delta(1 - \alpha)^2 v_h$$

where v_i is the continuation value given that you have observed outcome i. The average value of defecting against σ in the first period and then cooperating, v_d, is

$$v_d = (1 - \delta)[e\alpha(1 - \alpha) + f(1 - \alpha)^2 + g\alpha^2 + h\alpha(1 - \alpha)] \\ + \delta\alpha^2 v_f + \delta\alpha(1 - \alpha)v_e + \delta\alpha^2 v_g + \delta\alpha(1 - \alpha)v_h$$

For cooperation to be an equilibrium, the payoff to cooperation must be higher than the payoff to defection, $v_c \geq v_d$. Since σ is symmetric, $v_f = v_g$. Symmetry also allows us to write continuation values as a linear combination of the maximum value, v_c, and the lowest conceivable value, h. We could always replace σ by a strategy $\tilde{\sigma}$ and replace continuation strategies by a probability that either the worst or the best possible continuation is chosen. The actual continuation after a given observation would then depend on the realization of the publicly observable random variable. Define $p(q)$ to be the probability of the worst continuation given that h (f or g) is observed. Then continuation values can reexpressed as

$$v_f = (1 - q)v_c + qh$$
$$v_h = (1 - p)v_c + ph$$

Note that the use of p and q is not essential to the calculation of the upper bound; although it is convenient in establishing existence of a strategy that attains the upper bound and for finding such a strategy. Observe that v_e has a larger coefficient in the decomposition of v_c than in the decomposition of v_d. Therefore, if $v_e < v_c$ we could increase v_e without violating incentive compatibility. Since increasing v_e increases v_c, we must have $v_e = v_c$. For convenience define

$$A = e\alpha^2 + f\alpha(1 - \alpha) + g\alpha(1 - \alpha) + h(1 - \alpha)$$

A is the expected one period payoff if everyone cooperates. B is the expected payoff if one country defects but the other still cooperates

$$B = e\alpha(1 - \alpha) + f(1 - \alpha)^2 + g\alpha^2 + h\alpha(1 - \alpha)$$

Then v_c and v_d can be written as

$$v_c = \frac{(1-\delta)A + 2\delta\alpha(1-\alpha)v_f + \delta(1-\alpha)^2 v^h}{1 - \delta\alpha^2}$$

$$v_d = \frac{(1-\delta)B + \delta(1-\alpha)^2 v_f + \delta(1-\alpha^2)v_f + \delta\alpha(1-\alpha)vh}{1 - \delta\alpha(1-\alpha)}$$

Next notice that any optimal strategy must have $v_c = v_d$ since if the inequality held strictly, the probability of punishment could be lowered raising the expected payoff without violating the incentive constraint. Using this fact equate v_c to v_d and substitute in for v_f and v_h to get p as a function of q, α, δ, e, f, g and h

$$p = \frac{(B-A)(1-\delta) + [(B-h)2\delta\alpha(1-\alpha) - (A-h)\delta((1-\alpha)^2 + \alpha^2)]q}{(A-h)\delta\alpha(1-\alpha) - (B-h)\delta(1-\alpha)^2}$$

Below we will argue that v_c is a strictly decreasing function of q. Therefore, unless this violates other conditions, we want to set $q = 0$. The other conditions to worry about concern p. Namely, the p implied by a choice of $q = 0$ must be a probability. We next show that if $q = 0$, then p is positive. If $q = 0$ the numerator is positive since $B - A = (2\alpha - 1)[(g - e)\alpha + (h - f)(1 - \alpha)] > 0$. The denominator is positive if $\alpha A - (1 - \alpha)B - h > 0$. This will hold for α close enough to one. Also, given $q = 0$, p will be small for δ close enough to 1.

Given $q = 0$ it follows that the optimal strategy results in a payoff smaller than the efficient outcome. First express v_c as a function of p and q.

$$v_c = h + \frac{(1-\delta)(A-h)}{1 - \delta + 2\delta\alpha(1-\alpha)q + \delta(1-\alpha)^2 p}$$

Now let $q = 0$

$$v_c = h\left(1 - \cfrac{1}{1 + \cfrac{(1-\alpha)(B-A)}{(A-h)\alpha - (B-h)(1-\alpha)}}\right) + A \cfrac{1}{1 + \cfrac{(1-\alpha)(B-A)}{(A-h)\alpha - (B-h)(1-\alpha)}}$$

Since v_c is a linear combination of h and A.

$$v_c < A$$

which establishes that given $q = 0$, the optimal symmetric payoff is bounded away from the efficient payoff, A; note that δ does not enter the expression for v_c if $q = 0$. The final step is to show that it is optimal to set $q = 0$. Using the above expression for v_c as a function of p and q we substitute for p to get v_c as a function of q

$$v_c = h + \cfrac{(1-\delta)(A-h)}{1 - \delta + 2\delta\alpha(1-\alpha)q + \delta(1-\alpha)^2}\left[\cfrac{(B-A)(1-\delta) + [(B-h)2\delta\alpha(1-\alpha) - (A-h)\delta(1-\alpha)^2 + \alpha^2]q}{(A-h)\delta\alpha(1-\alpha) - (B-h)\delta(1-\alpha)^2}\right]$$

Taking the derivative with respect to q we find that it is negative if $\alpha > 1/2$. Therefore the optimal q is zero. This means that the optimal strategy involves punishing only when the (h, h) outcome is realized. This is in sharp contrast to the result when $\alpha = 1$ which is that punishment occurs when an off-diagonal outcome is observed.

The intuition for this result is that with punishment "on the diagonal," the likelihood of triggering punishment when there was indeed a defection divided by the likelihood of triggering punishment when there was no defection equals

$$\cfrac{\alpha}{1-\alpha}$$

whereas the same ratio equals

$$\cfrac{2\alpha^2 + 1 - 2\alpha}{2\alpha(1-\alpha)}$$

for punishment "off the diagonal." The latter ratio is smaller for $\alpha > 1/2$ which means that the event "lower-right-hand corner" is more informative about whether there was a defection than the event "off the diagonal."

It follows from the above that punishing forever with some probability will support the best payoff attainable with symmetric strategies. Furthermore, despite the folk theorem this best payoff will not be the efficient one.

7. Asymmetric Strategies Revisited

We saw in the previous section that average payoffs from symmetric strategies are bounded away from the efficient frontier and that there is an upper bound which is independent of the discount factor. On the other hand, we know from the Folk theorem that for a sequence of discount factor converging to 1 there must be an associated sequence of strategy profiles while form perfect public equilibria and whose payoffs converge to the maximal symmetric payoff. Our goal in this section is to assess how difficult it is to construct such a sequence. For that purpose we will as in the previous section make use of a public randomizing device. This should facilitate the construction of the desired sequence; recall that in the previous section the randomizing device allowed us to construct simple strategies which obtained the highest payoff available from symmetric strategies. Since the computations are tedious we will go back to our numerical example of the prisoners' dilemma. This is not an essential limitation because our result will be negative. We will show that within a large class of simple asymmetric strategies it is impossible to construct a sequence of perfect public equilibrium strategies whose payoffs converge to the maximal symmetric payoff of the stage game as the discount factor converges to 1.

Let there be a public randomizing device which implements a probability of punishment p and a probability of returning to cooperation q, i.e., with probability p an event "punish" will occur and with probability q an event "return" will occur. Let history be summarized by one of three states, C, Q_1 and Q_2, a cooperative state, a punishment state for player 1 and a punishment state for player 2 and the current realization of the public randomizing device. Consider the following strategy for the row player: In state C the strategy prescribes U, in state Q_1 the strategy prescribes U and in state Q_2 the strategy prescribes D. The state transition rule is as follows. In state C return to C after (U, L) and (D, R); after (D, L) go to Q_1 if the event "punish" occurred, otherwise stay at C; after (U, R) go to Q_2 if the event "punish" occurred, otherwise stay at C. In Q_1, stay at Q_1 after (D, R), otherwise return to C if the event "return" occurred and stay at Q_1 if that event did not occur. In Q_2, stay at Q_2 after (D, R), otherwise return to C if the event "return" occurred and stay at Q_2 if that event did not occur. Suppose the column player uses the same strategy, except that in Q_1 the column player uses R and in Q_2 the column player uses L.

By varying p and q, the above construction traces out a large class of strategy profiles despite the restriction to only three states. In particular, as we will see, one can use the probabilities to remove slack from incentive constraints. If there are non-binding incentive constraints, the probabilities can be adjusted to increase payoffs without violating any of the incentive constraints. Overall, the probabilities can be chosen such that punishments are severe but not invoked more frequently than necessary to force agents to conform.

Denote by u player one's expected payoff in state C from both players conforming with the above profile. Let v and w be player one's payoffs from conforming in states Q_1 and Q_2 respectively. Let x stand for player one's payoff from defecting once in state C and conforming thereafter; let y be her payoff from defecting once in state Q_1 and conforming thereafter. One can ignore possible defections of player one against player two's punishment because such defections would lower expected payoffs in that stage and also would reduce the possibility of remaining in player two's punishment state which is clearly the most advantageous state for player 1.

u, v, w, x and y must satisfy the following equations:

$$u = (1 - \delta)[5\alpha^2 + 7\alpha(1 - \alpha) + (1 - \alpha)^2]$$
$$+ \delta[\alpha^2 u + \alpha(1 - \alpha)[pv + (1 - p)u]$$
$$+ \alpha(1 - \alpha)[pw + (1 - p)u] + (1 - \alpha)^2 u]$$
$$v = (1 - \delta)[6\alpha(1 - \alpha) + 7(1 - \alpha)^2]$$
$$+ \delta[\alpha(1 - \alpha)v + (1 - \alpha(1 - \alpha))[qu + (1 - q)v]]$$
$$w = (1 - \delta)[7\alpha^2 + 6\alpha(1 - \alpha)]$$
$$+ \delta[\alpha(1 - \alpha)w + (1 - \alpha(1 - \alpha))[qu + (1 - q)w]]$$
$$x = (1 - \delta)[7\alpha^2 + 6\alpha(1 - \alpha)]$$
$$+ \delta[\alpha^2[pv + (1 - p)u] + 2\alpha(1 - \alpha)u + (1 - \alpha)^2[pw + (1 - p)u]]$$
$$y = (1 - \delta)[\alpha^2 + 5(1 - \alpha)^2 + 7\alpha(1 - \alpha)]$$
$$+ \delta[\alpha^2 v + (1 - \alpha)^2[qu + (1 - q)v]]$$

From the first of these expressions it follows that $2u > v + w$; otherwise we would have $u > 5\alpha^2 + 7\alpha(1 - \alpha) + (1 - \alpha^2)$ which is impossible.

The relevant incentive constraints for player one are

$$u \geq x \text{ and } v \geq y.$$

It turns out that if we are interested in optimal perfect public equilibria, then it is without loss of generality to require that both of these constraints hold as equalities. To see this, suppose that we have a perfect public equilibrium in which $v > y$.

Now consider reducing q. One can show that for u constant

$$\frac{\partial(v + w)}{\partial q}\bigg|_{u - \bar{u}} > 0.$$

Thus, lowering q lowers $v + w$, conditional on holding u constant; since $v + w < 2u$, u can be held constant by an appropriate choice of p. Hence, for any equilibrium payoff u, if in the supporting equilibrium the constraint $v \geq y$ has slack, we can find another equilibrium in which q is lower and which supports the same u. Since for $q = 0$ we would have $y > v$ there must be a lower bound on q to remain consistent with incentive compatibility. Hence, it is without loss of generality to let $v = y$. Suppose now that $u > x$. Lowering p will increase u since $v + w < 2u$. At the same time increasing u will loosen the constraint $v \geq y$. Thus, whenever $u > x$, p and therefore payoffs can be increased without violating incentives. Hence it is without loss of generality to let both

$$u = x \text{ and } v = y$$

when we are looking for maximal symmetric payoffs. This gives us a system of seven equations in seven unknowns.

By solving the above system of equations, one obtains

$$u = \frac{\sqrt{4\alpha^6 - 56\alpha^5 + 164\alpha^4 - 136\alpha^3 + 64\alpha^2 - 16\alpha + 1} + 2\alpha^3 + 8\alpha^2 - 4\alpha - 1}{2(2\alpha - 1)}$$

Observe that this expression is independent of the discount factor. This fact makes it easy to compare u with the efficient symmetric payoff in the stage game, $f = 5\alpha^2 + 7\alpha(1 - \alpha) + (1 - \alpha)^2$. One can show that $f - u$ is strictly positive for all α in the interval $(1/2, 1)$. Therefore, for all relevant α, within the class of strategies considered it is impossible to find a sequence of perfect public equilibria with payoffs converging to efficient payoffs. More complex strategies are needed if one wants to obtain approximately efficient payoffs.

Appendix

In this appendix we sketch an argument to show that the upper bound on payoffs from symmetric perfect public equilibria (SPPE) is attained by an SPPE.

Denote by v_c the supremum of the set of payoffs from SPPEa. Consider a sequence of SPPEa with an associated payoff sequence $\{v_c^n\}_{n=1}^{\infty}$, $v_c^n > v_c^{n-1}$, $v_c^n \to v_c$ and continuation payoffs v_e^n, v_f^n, v_h^n, following the realizations e, f, and h. Note that payoffs are bounded and can be chosen to lie in a compact set such that the assumed convergence is no problem. By choosing further subsequences if necessary it follows that there exists a sequence of SPPEa such that v_j^n converges to some v_j for all $j = e, f, h$. For the n-th strategy in the sequence denote by v_d^n the payoff from defecting in the first period and then conforming and note that v_d^n converges since it is a sum of converging terms. For $j = e, f, h$ define p_j^n as the solution to

$$v_j^n = (1 - p_j^n)v_c^n + p_j^n h.$$

Note that p_j^n converges by construction and the limiting values p_j satisfy $p_j \in [0, 1]$. Also, the inequality $v_c^n \geq v_d^n$ is preserved in the limit such that $v_c \geq v_d$. Now construct a strategy which uses two states indexed by c and h. The c-state is the initial state and the strategy prescribes cooperation in this state. In the h-state the strategy prescribes defection. The h-state is absorbing. If the c-state is the current state, the state transition rule is that after a j-event the new state is the h-state with probability p_j and the c-state otherwise. It is easily verified that if both players follow this strategy (with relabeling for the column player), then each players expected payoff equals v_c and the payoff from a one-time defection in the c-state equals v_d. Since there are only two states and the continuation in the absorbing state trivially is an equilibrium, the only relevant incentive constraint is $v_c \geq v_d$ which is satisfied as argued above.

Notes

1. A closed and convex subset of $\mathbf{R}^n$ is called *smooth* if it has non-empty interior and its boundary is a C^2-submanifold of $\mathbf{R}^n$.

2. It might in fact be much easier, not to say trivial, to get the folk theorem in a full blown economic model. The problem which we are concerned with is that for any tariff policy of one country the other may have a tariff policy which induces the same distribution over observables irrespective of the status quo. The intuition which led us to think of this as a problem was that from the status quo there are always two ways to induce any given t.o.t. with at most one country changing its tariff. If the errors are just linearly tagged on to this, this intuition carries over to the stochastic case. However, whether this will be true in general, hinges on the error structure. And it seems very plausible that if the errors are located somewhere "inside the model," then it would indeed be very unlikely for any two pairs of tariffs to induce the same distribution. Once this link is broken, there is no obvious reason for the Folk Theorem not to hold.

References

Abreu, D. 1988. "On the Theory of Infinitely Repeated Games with Discounting." *Econometrica* 56: 383–396.

Abreu, D., D. Pearce, and E. Stacchetti. 1986. "Optimal Cartel Equilibria with Imperfect Monitoring." *Journal of Economic Theory* 39: 251–269.

Abreu, D., D. Pearce, and E. Stacchetti. 1990. "Toward a Theory of Discounted Repeated Games with Imperfect Monitoring." *Econometrica* 58: 1041–1063.

Abreu, D., and A. Rubinstein. 1988. "The Structure of Nash Equilibrium in Repeated Games with Finite Automata." *Econometrica* 56: 1259–1281.

Dixit, A. 1987. "Strategic Aspects of Trade Policy." In *Advances in Economic Theory Fifth World Congress*, Truman Bewley (ed.) New York: Cambridge University Press 329–362.

Fudenberg, D., and E. Maskin. 1986. "The Folk Theorem in Repeated Games and with Incomplete Information." *Econometrica* 54: 533–554.

Fudenberg, D., D. Levine, and E. Maskin. 1991. "The Folk Theorem with Imperfect Public Information." Working Paper.

Green, E., and R. Porter. 1984. "Noncooperative Collusion under Imperfect Price Information." *Econometrica* 52: 87–100.

Kennan, J., and R. Riezman. 1988. "Do Big Countries Win Tariff Wars?" *International Economic Review* 29: 81–85.

Matsushima, H. 1989. "Efficiency in Repeated Games with Imperfect Monitoring." *Journal of Economic Theory* 48: 428–442.

Radner, R., R. Myerson, and E. Maskin. 1986. "An Example of a Repeated Partnership Game with Discounting and with Uniformly Inefficient Equilibria." *Review of Economic Studies* 53: 59–70.

Riezman, R. 1991. "Dynamic Tariffs with Asymmetric Information." *Journal of International Economics* 30: 267–284.

Rubinstein, A. 1979 "Equilibrium in Supergames with the Overtaking Criterion." *Journal of Economic Theory* 21: 1–9.

10 COORDINATION THEORY, THE STAG HUNT AND MACROECONOMICS

John Bryant

1. Introduction

Coordination theory may well have application to systems generally. In particular, coordination theory may provide insights for systems in which a diversity of outcomes seems to characterize seemingly similar environments, for system instability, and for decentralized system collapse. Coordination theory examines situations involving multiple, and, in the extreme, a continuum, of Nash equilibria, and focuses on the particularly striking cases in which the Nash equilibria are Pareto ranked. Possible applications of coordination theory include, but are not limited to, physical and engineering systems, and biological, medical, behavioral, ecological and evolutionary systems, as well as social, political and historical systems. This paper has, however, a narrowed focus, and a more modest aim of considering a few possible applications of coordination theory to economics, particularly Macroeconomics. The basic theme is that coordination theory may eventually provide a piece of one possible approach to some aspects of Macroeconomic behavior. No attempt is made to be comprehensive or even-handed in defending this theme. Rather, notions of potential applications of coordination theory, that have happened to strike the author's fancy, are presented. For further reading, the author recommends Roberts (1987), Mankiw and Romer (1991), and Guesnerie (1992). Early sources include Hayek (O'Driscoll, c. 1977) and Keynes (1936).

2. Background

Coordination theory treats integrated systems of economic agents. While it may be more matter of shades of gray than of black and white, nonetheless, this systems orientation of coordination theory may prove to be what most clearly distinguishes a coordination theory approach to Macroeconomics from more traditional approaches. For example, Colander (1992) makes this point. In particular, Colander emphasizes that the distinguishing feature of the coordination theory approach to Macroeconomics is its stress on the interdependence of individual's decisions and expectations, and its denial that "the aggregate economy can be analyzed by analyzing the individual's decision independent of the interaction with the whole." (Colander, 1992, p. 8). Individual's decisions are interdependent, and they recognize this interdependence. In brief, behavior is social.

In traditional game theory, "solution concepts" bridge the gap between Robinson Crusoe and social behavior of an integrated economic system. The standard model of behavior used in economics, utility theory, treats only simple individual choice (explicitly, anyway). However, if the term "economy" means anything, and certainly if the term "Macroeconomy" means anything, social behavior must be at the ultimate basis of the discipline. In game theory, it is the "solution concept" that moves simple individual choice into the realm of social behavior. Ultimately, however, it may prove useful to proceed to a more primitive level than "solution concept" in order to treat integrated economic systems. One wants to be at the level of assumptions on the physical environment and on individual psychology, and the implications thereof.

One can only speculate at the consequences when, and if, the analysis in game theory moves to a more primitive level than the "solution concept." One intriguing possibility is that it may turn out the tastes, endowments and technologies are not a complete model of a integrated economic system. That is to say, it may be necessary to specify more about the environment before behavior can be inferred. In particular, behavior in simple choice settings may have only limited implications for social behavior. This possibility may, in turn, raise some alternative scenarios for future research programs. One further possibility is that it will prove useful to specify a richer characterization of the individual psyche than just a utility function (or the behavior in simple choice situations implied by a utility function). An alternative further possibility is that implications of tastes, endowments and technologies, while limited, are robust. It may be that tastes, endowments and technologies put limits on behavior that are widely followed, and fairly readily computed and predicted. Further, it may be that the details of behavior, within those limits, depend upon complicated features of the individual psyche, or of the physical environment, or of both interacting, and which are hard to observe or predict. Under this scenario, economics may have been trying, and failing, to predict precise behavior in the face of strong, predictable regularities in some of the behavior not engaged in. It may be hard to predict what is done, but easy to predict some of what is not done. An analogy might be that, if you are trying to find a skater, knowing the location of the rink gives you a good start, while predicting precisely where in the rink is a different order of problem! Approaches like iterated contraction by dominance, for example, are suggestive of this. In any case, coordination theory, when full developed, may, in some circumstances, provide a piece of the more primitive bridge between Robinson Crusoe and the social behavior of an integrated economic system.

Perhaps Walrasian equilibrium might be characterized as *the* traditional approach to Macroeconomics. Walrasian equilibrium may abstract from some integrated systems aspects of the economy. In Walrasian equilibrium, economic agents might appear more as semi-autonomous Robinson Crusoes, than as elements in an integrated economic system. Individuals are faced with simple choices, and social behavior is subsumed in the (supposedly) impersonal market forces behind the prices of the invisible hand. However, Walrasian equilibrium is an "as if" model of the market (if it is coherent). It is not a mechanism. Exactly where do these hyperplanes, arching across the sky, as it were, come from? Not only then, in reality, may there be economically important social behavior that occurs outside "the market" (somehow defined), the market itself may involve social behavior from which Walrasian equilibrium abstracts. For many purposes, abstracting away from non-market interaction, and from real market mechanisms, may be very useful and practical. However, for some purposes, and in some circumstances, the finesse of the gap between Robinson Crusoe and the social behavior of an integrated economic system, which Walrasian equilibrium may appear to be, may abstract away from interesting economic phenomena.

A third traditional approach to macroeconomic behavior is provided by Aggregative Macroeconomic Theory. Many would trace this approach to the work of Keynes in his *General Theory*, and it is the approach advocated by James Tobin (e.g., 1982) in his program for Macroeconomics. Aggregative Macroeconomic Theory sidesteps the gap between Robinson Crusoe and the social behavior of a integrated economic system. In modeling at the aggregate level, Aggregative Macroeconomic Theory simply has no gap to address. The cost, of course, is that Aggregative Macroeconomic Theory lacks microeconomic underpinnings. Admittedly, many aggregative macroeconomic models seem to have their structures justified by loose allusions to the underlying environment. However, it is hard to interpret the role of these loose allusions, as the gap between Robinson Crusoe and the social behavior of an integrated economic system, or, more generally, between the physical environment and the posited aggregate relationships, is not addressed in any formal, rigorous way. Perhaps the best way to interpret Aggregative Macroeconomic Theory is just to take the posited aggregate relationships as primitives. In any case, the lack of formal underpinnings is a substantial cost. For one thing, the Aggregative Macroeconomic Theory cannot, as a consequence, address normative issues in any formal way. Part of the very reason for attempting macroeconomic study is sacrificed from the start. There are negative repercussions for addressing positive issues as well. In particular, by structure, Aggregative Macroeconomic Theory cannot formally address the consequences for aggregate relationships of any observed changes to the underlying environment. Moreover, if those changes to the underlying environment have consequences for the Macroeconomy which are not captured in posited aggregate relationships, these consequences are simply lost to the theory. Whatever the short run practicality of the Aggregative Macroeconomic Theory approach, ultimately it may prove useful to proceed to a more primitive level in order to treat integrated economic systems.

One might speculate that there is one particular consequence of the systems orientation of coordination theory that may ultimately prove to distinguish coordination theory from both the traditional Walrasian equilibrium and the Aggregative Macroeconomic Theory approaches in a singularly important way. In both the Walrasian equilibrium and Aggregative Macroeconomic Theory approaches, it is the *presumption* that it is the role of the "price system" to coordinate activity. In both approaches, the mechanism of the coordination is not modeled, but abstracted away from, as discussed above. In Walrasian equilibrium, it may seem to be the *presumption* that this coordination is perfectly obtained. In some branches of Aggregative Macroeconomic Theory, it may seem to be the presumption that this coordination is imperfect, because of some failing in the price system. Since, when it comes down to it, the difference may prove to be one of presumption, of different primitives, the long debate on this issue may prove to have generated more heat than light (but perhaps also some well paid careers in the meantime). Coordination Theory, on the other hand, suggests the possibility that, in some circumstances, entrepreneurs and institutions, more than the "price system," coordinate agents' activities in an integrated economic system. Non-market allocations are important. This further immediately suggests that any failure in coordination may not be due to failure in the "price system." In particular, price rigidity may not be the culprit in an observed failure in coordination. In standard Macroeconomic terms, fully flexible prices do not by themselves lead to optimal output (another point emphasized by Colander, 1992).

3. The Stag Hunt

It may be useful to put more structure on the discussion. An examination of the concrete example of the stag hunt production game can provide additional structure. Indeed, the stage hunt production game provides a particularly graphic parable of the Macroeconomy as an integrated economic system.

Colander (1986) provides a textbook formulation of the coordination theory approach to Macroeconomics. In particular, Colander specifies an aggregate production function of the form $Q = f(K, L; C)$ where Q is output, K is capital, L is labor, and C is the new variable reflecting degree of coordination in the economy. While this sort of formulation is doubtless pedagogically useful, to understand the role of the coordination parameter it may be helpful to proceed to a deeper level. The stag hunt is an easily understood example of an underlying structure which yields a continuous coordination parameter.

The stag hunt production game can be treated as a strategic game as in Van Huyck, Battalio and Beil (1990). For the purposes of this discussion of coordination theory and Macroeconomics it is more insightful to consider a version with more explicit economic content. In particular, the team production problem puts some economic meat on the stag hunt bones.

A simple version of the stag hunt production game (drawing on Bryant (1983)) is easily described. There are two consumption goods in the economy, leisure (C_1) and a commodity (C_2), multigrain bread. Every individual has an identical increasing and concave utility function $U(C_1, C_2)$, and has an endowment of L units of leisure. Bread is produced from leisure in a two stage process in which grain is produced from leisure in the first stage and bread is produced from grain in the second stage. The first stage of production occurs in the familiar island model. Each of N islands contains n people and each island produces a distinct grain. Each of the N grains must be used in precisely the same proportion to produce bread. One unit of grain is produced by one unit of leisure and one unit of bread is produced by N units of grain, one unit of each type. Thus each individual must decide how many units of leisure to devote to grain production and, letting G_i denote the total grain production in island i, bread output is min $\{G_1, \cdots, G_N\}$. Intuitively speaking, this is a world where no matter how many people are in it, people, even in the limit, never are small. By assumption, in the first stage of production, there is no communication, travel, transportation, no contact whatever, between islands. In the second stage of production there are no limits whatever to contact between islands. Effectively, there is a single production site in the second stage of production.

This model of the stage hunt production game economy exhibits two conceptually separate, but interrelated, problems, an allocation problem and a coordination problem. First we consider the allocation problem, which, for a particular specification of its resolution, generates the sort of coordination problem which is our main topic of inquiry. As it stands, the model has no allocation mechanism. To confront the allocation problem, consider first the traditional Walrasian equilibrium approach to the market. As there is no contact between the islands in the first stage of production, there is, by construction, no market for the commodity in the first stage of production. Consider the second stage of production. Walrasian prices do not work well. Suppose $G_1 < G_i$ for $i \geq 2$. Let P_i denote the bread price of grain. Then we must have $P_1 = 1$ and $P_2 = \cdots = P_N = 0$ due to the respective marginal products of the inputs, grains. At any positive price, all grain is supplied to the market, it is useless for anything else, and "all" is an excess supply for grains 2 through N. $P_1 = 1$ exhausts the supply of bread, and is the marginal product of grain one. 0 is the marginal product of grains 2 through N. If two or more grains are scarce, prices are indeterminate. For example, if grains 1 and 2 alone are scarce, then any set of prices that satisfies $P_1 + P_2 = 1$, $P_3 = \cdots = P_N = 0$ is an equilibrium. The marginal product of grains one and two are indeterminate in this case, 0 up and 1 down. If we consider the "downwards" marginal product, the sum of the marginals exceeds the total. This can be taken to be the definition of a team production technology. One way to view this problem is that there are innately unowned rents. There is, if you will, an unowned scarce "factor," the right to get together, and, in fact, this scarce factor accounts for all the rents in this example.

One could ignore the indeterminacy of Walrasian prices in the stag hunt production game and impose the convention that scarce factors face the same price; although it is not clear whether and how markets could actually do this. However, under this scenario, if the individuals understand the process generating prices, the first stage production problem is a continuous version of the Prisoners' Dilemma, between islands. That is, since the scarce island gets the entire product, the Nash equilibrium is zero output. If the goal of the example were simply to demonstrate "pathology," in the Walrasian point of view, from a distinctly non-pathological, almost work-a-day, environment, that is achieved. However, the goal is to explore specifically the need for coordination, as distinct from cooperation.

Consider, instead, what may be a more reasonable approach to the allocation problem, in this context of the stag hunt production game. Imagine that at each island, separately, in the first stage of production, individuals somehow reach agreement, and produce equal amounts of their preassigned grain. Imagine, further, that, in the second stage of production, each island, separately, somehow agrees to, equally for each individual, destroy any excess grain (this story is not really necessary, the author just finds it appealing). Then observing the symmetry of the situation, the largest possible amount of bread is produced, and divided equally between all individuals. This scenario of equal division is consistent with standard models of bargaining. It may also be worth noting that equal division may seem an even more natural outcome in a team problem in which input is not observable, only joint output. Admittedly this may be a somewhat Pollyannish view of the market. The difficulty we have had resolving this allocation problem is, itself, a manifestation of the gap between Robinson Crusoe and the social behavior of an integrated economic system. The coordination problem, to which we now turn, is another, seemingly simpler, manifestation of this gap.

This specification of the resolution of the allocation problem, equal division, generates the sort of coordination problem which is our main topic of inquiry. Suppose the individuals in the first stage of production correctly anticipate the equal division in the second stage of production. Let G^* be the optimal individual amount of grain produced. Then any G satisfying $O \leq G \leq G^*$ is a Nash equilibrium level of grain production. This continuum of Nash equilibria are Pareto ranked, with G^* Pareto optimal. This is what Van Huyck, Battalio and Beil (1990) refer to as to as a pure coordination game. One way to think about the result is in terms of prices. Suppose all the individuals at island one, say, share the belief that the minimum per individual grain production at the other islands will be $\tilde{G}$,

$O \leq \tilde{G} \leq G^*$. Then, for the first $\tilde{G}$ units per individual of grain one, these individuals expect to receive "price" $P_1 = 1/N$, and for all units in excess of $\tilde{G}$, zero. As $\tilde{G} \leq G^*$, the individuals at island one, with this belief, produce $G = \tilde{G}$.[1]

Crucial ingredients in the stag hunt production game are specialization and imperfect information. Individuals are specialized in what they produce, and interconnected in what they consume, and are, hence, interdependent.

The imperfect information feature of the stag hunt production game requires more discussion. In this model, communication is impossible in the first stage of production, and communication is perfect and costless in the second stage of production. There is a natural tension in an economic coordination game, namely, if individuals can affect each other, and hence, in some sense, must be making contact, why, then, cannot they communicate? What one has in mind is that at the time an investment decision is made, it is just not practical even to identify, much less contact, all those individuals who ultimately might affect the return on that investment. For example, when one introduces new Coke, one does not contact all potential future purchasers of the product, and then get them to sign appropriate contingent contracts to buy, and drink the stuff. Similarly, when one decides whether or not to put a gas station at a particular corner, one does not contact all potential future purchasers of the gas. So what seems a little awkward in modeling, may be perfectly natural in reality.

It also may be worth stressing that imperfect information might more usefully be thought of as an inability to write the appropriate multilateral contracts in timely fashion. Indeed, the previous discussion suggests this interpretation. Thus it appears that it is multilateral contracting which is required to eliminate the coordination problem in the stag hunt production game, and not, in particular, the sets of bilateral contracts observed on actual centralized exchanges. (The author is indebted to Robert Solow for stressing the importance of this point to him.)

As something of an aside, the author also notes that such information costs may provide a practical argument against the use of Walrasian equilibrium in some contexts. Typically Walrasian equilibrium is thought to be most nearly applicable in situations in which the number of agents is large. However, as the number of agents in the economy goes to infinity, the per capita cost of the assumed communication could well go to infinity as well. Keep in mind that a minute's advertising on the Super Bowl costs on the order of a million bucks, and this allows only one way communication, not bilateral communication, much less bilateral contracts, much less multilateral communication, much less multilateral contracts.

4. Some Potential Applications

Several features of the stag hunt production game are of potential Macroeconomic interest.

The stag hunt production game models effective demand and involuntary underemployment. This point has been stressed by Roberts (1987) and Phelps (1990). The model illustrates that even given "competitive" prices, the recognition of quantity constraints has the potential to generate inefficiency and underutilization. It illustrates that prices may not be the problem. Roughly speaking, if there is a coordination problem, and there is something you might try to do about it, why work on prices, when it is the quantities that are the problem. Work on the quantities, the prices are just fine. While in a general equilibrium model supply and demand are not easily separated, in the stag hunt production game in an inefficient equilibrium suppliers of intermediate good (grain) are anticipating low levels of sales of their product. At the same time, because of this low anticipated level of sales, individuals input less of their leisure than they would like. This is a simple version of effective demand and involuntary unemployment. Roberts (1987) has initiated an exciting research program of providing a more complex industrial organization and market structure in which effective demand and involuntary unemployment are manifested in more elaborate ways. The models are quite complicated, and more work needs to be done on this genre before it can be evaluated (see e.g., Jones and Manuelli (1992)). Perhaps it will prove helpful to this evaluation if more of the industrial organization and market structure can be made endogenous.[2] It is worth pointing out that while Roberts' approach involves individuals who are specialized in what they produce and interconnected in what they consume, like the stag hunt production game, it does not involve the technological complimentaries of the team production of the stag hunt production game. It remains to be seen whether technological complimentaries are an important factor in their own right in generating effective demand and involuntary unemployment.

The stag hunt production game may capture the idea of a rigid world. Through the years many writers have expressed the belief that Keynes had in mind some sort of a "rigid" world, where it is hard to get things to fit together, whereas the classicals had in mind a fluid world, where things just flow together (see e.g., Milo Keynes, 1975). The fixed proportions production technology may exhibit the ultimate in "rigidity". Phelps (1991), and others, have argued that "rigidity" is a misinterpretation of Keynes, and really the contribution of earlier writers. Note, however, in this regard, that the rigidity generated in the stag hunt production game is not nominal price rigidity, but a basic technological rigidity. Things have to fit together right in this world. One might also note in this regard that this sort of rigid world view of the Macroeconomy may, in some contexts, be more relevant for short run rather than long run behavior. The ability to redesign technologies in the long run may reduce rigidities. (Note also that this rigidity is not necessary for the central story of underemployment equilibria, although it may be of substantial independent interest.)

The stag hunt production game suggests that kinks may have a role to play in explaining a continuum of Macroeconomic equilibria. When one restricts attention to smooth technologies, it appears that generically there are a finite number of equilibria (see Cooper and John, 1988). However, kinks tend to produce continua of equilibria. It is hard to know what to take as the underlying natural measure, but one might speculate that generically technologies are not smooth! More down to earth, kinks may be as much a defining characteristic of team production as is the sum of the marginals exceeding the total product.

The stag hunt production game isolates a role of strategic uncertainty in the Macroeconomy. As Van Huyck, Battalio and Beil (1990) note, the stag hunt is a pure coordination game, and as they and Crawford (1991) stress, all that may be going on in the stag hunt is strategic uncertainty. At the same time, the work-a-day nature of the team production problem suggests, then, that strategic uncertainty may be a pervasive and important phenomenon in the economy. As shown by Van Huyck, Battalio and Beil and by Crawford, there are no conflicting interests or moral hazard aspects to the stag hunt. Hence, the team production problem may isolate the simplest of social problems of an integrated economic system. Perhaps, then, examination of the stag hunt production game may be the logical first step in bridging the gap between Robinson Crusoe and the social behavior of an integrated economic system.

The stag hunt production game suggests several possible issues to be addressed by a coordination theory approach to the Macroeconomy as an integrated economic system. In fact, whole hosts of such possible issues just pop to mind. What follows, then, is a sort of laundry list of some of the issues which happen to have popped into the author's mind.

The stag hunt production game may be a particularly graphic parable, or, alternatively, a real structure characterizing some aspects of the Macroeconomy. The stag hunt production game presents perfect complements, which makes it sharp and graphic, and, hence, possibly, of didactic value. But there is also the possibility that this very structure is out there. It might even be that this very structure presents particularly severe difficulties in pure coordination, so that identifying its presence is particularly important.

Typically in Macroeconomic coordination models it is assumed that there are global externalities. Indeed Colander (1992) considers macro externalities to be the keystone of coordination Macroeconomics, and, in particular, of the distinguishing feature of the approach, namely the denial of "the independence assumption: [that] the aggregate economy can be analyzed by analyzing the individual's decision independent of interaction with the whole" (p. 8). The economy, at least in some aspect, comprises a single team, or all agents are simultaneously interconnected in some manner. There is another possibility that can generate an economy wide coordination problem, however. The Macroeconomy may exhibit aspects of the interlocking complementarity of overlapping teams. It may be that in some aspects, and to some degree, that (a) has to fit with (b), which has to fit with (c), which—, which has to fit with (a). Examples of interlocking complementarity include Bryant (1984) and Durlauf (1989), which draw on Samuelson's familiar overlapping generations model. These models, to date, are fairly abstract, and could use more real economic meat.

The stag hunt production game may help isolate those structures which are amenable to coordination, and those which are not. It may be that the stag hunt is a particularly difficult pure coordination problem, and hence characterizes situations in which either coordination is not achieved, or in which substantial institutional development is required for coordination. On the other hand, as pure coordination games, stag hunt structures may be amenable to coordination in ways that structures also involving conflicting interests or moral hazard aspects are not. Moreover, variants on the simple stag hunt may be easier or harder to coordinate, and may characterize various real structures, and serve to help identify them.

The stag hunt production game may help clarify the role of institutions, that is firms, banks, trade associations, governments, and such, in coordination. Of course, in some approaches to economics, it is difficult to explain the very existence of institutions at all, or, at least, when institutions exist in a model, they frequently play no important role. Coordination may be an important role of institutions. Moreover, different variants of the coordination problem may encourage the formation different kinds of institutional structures. For example, the stag hunt may require multilateral contracts, or a firm structure, for solution, while less difficult pure coordination games may be solvable by exchanges, or by loose trade associations.

The stag hunt production game may help clarify the role of exogenous shocks in the Macroeconomy. Perhaps shocks sometimes change the structure of the economy so that it goes from being a game that does not have a coordination problem to a coordination game, or alters the variant of the coordination problem and, hence, the sort of institutions that are appropriate for coordination. In this way, shocks might have impacts that seem disproportionate to their size. Perhaps, also, shocks introduce uncertainty to the system, additionally, reintroducing strategic uncertainty that had previously been resolved, and, consequently, once again, having a disproportionate impact.

The stag hunt production game may help explain why turning points are so hard to predict. The character of the game, to a large extent, depends upon out of equilibrium, off diagonal, payoffs. It well could be that changes occur in the economy which influence out of equilibrium payoffs, but have little or no effect on equilibrium payoffs. Then the game could be altered from one without a coordination problem to one with a coordination problem, without affecting any observed returns. Then, out of the blue, as it were, the economy goes to hell in a hand basket. In a recovery, the reverse could happen.

Indeed, there are many more economic issues for which coordination theory may provide some insights. Possibly coordination theory may help explain endogenously generated shocks. For the nineteenth century business cycle theorist, unlike the modern real business cycle theorist, the whole mystery of the business cycle was why the economy went into recession, without an observable external shock. A clear possibility is that shocks are endogenous to the system. In the coordination game, strategic uncertainty is endogenous to the system. On the other hand, coordination theory may help explain persistence. Once the economy gets to a bad equilibrium, somehow, it stays there until there is some structural improvement. Coordination theory may help explain the business cycle. For example, perhaps there is a fixed cost to setting up and maintaining coordination institutions. Only when the economy is in deep recession is it worthwhile to coordinate. Once this coordination is achieved, the coordination institutions are allowed to deteriorate. Eventually the economy fails to coordinate once again, and the cycle repeats itself. Perhaps different teams, or team aspects, in the economy hit those critical points at different times, and at different levels of economic activity. Of course, in this particular example, there is the positive question of whether one observes coordination institutions being formed when the economy is depressed. Perhaps, also, coordination theory may have relevance to some aspects of the Great Depression, or of bank runs. Perhaps coordination theory has relevance to development. In this regard, one small, very speculative, point is that we may live in a sort of fractal world. As production and information technologies advance, larger and larger teams are possible, but there is always a margin where information is costly to convey perfectly and coordination problems occur—but the potential return to the large team may justify the "costs" of miscoordination. Perhaps coordination theory has relevance to the "Soviet Union" economic implosion. Indeed, possibly centrally planned economies encourage exactly those technologies which are most subject to coordination problems, when the planner is removed! More generally, coordination theory may help isolate those structures which are amenable to decentralization, and how certain structures can be usefully decentralized. Perhaps coordination theory will help illuminate the role of communication. In a world in which the only role of communication is to rip off the unwary, it is hard to explain the development of language, much less the huge expenses we observe on communications technology in advanced economies. Perhaps, more generally, coordination theory is one basis of normative economics. There may be some hope, I suppose, that economics might help move a society from a bad equilibrium to a good one.

There is also a question of the level in a society at which coordination is achieved. Coordination may occur at the level of explicit institutions, as discussed above, or at the level of constitutions and legal systems (some of my Chicago friends are willing to go so far as to say that the United States coordinated on property rights, while some other societies coordinated on bad equilibria). Or, perhaps, coordination may occur at the level of values.

5. Experimental Stag Hunt

Van Huyck, Battalio, and Beil (1990) (VHBB) experimentally tested the stag hunt production game. A fuller presentation is available in VHBB and in Crawford (1991). VHBB used a strategic version of the game in which there were several discrete choices, and a worst choice. The author agrees with Crawford that these experiments were remarkable for the intelligence and care which went into their planning and execution, both in concept and in practice. The author also shares the speculation that this remarkable effort may account for the striking clarity of the experimental results obtained. These may be the measure of future experiments; in short, these are dynamite experiments! Related exciting experiments are presented in VHBB (1991) and Cooper et al. (1990, 1992).

For larger groups, VHBB found that in the first play actions were widely dispersed. Under repetition with the same group of subjects, the actions converged systematically to the worst equilibrium, whether or not in the first play any subjects played the corresponding "worst" strategy (the latter is termed "overshooting"). VHBB found many other striking regularities, possibly quite important, which they and Crawford discuss.

VHBB's experimental results with the stag hunt production game raise several more issues to be addressed by a coordination theory approach to the macroeconomy. Is the first play of the game relevant and, hence, disequilibrium important in the macroeconomy? If so, is disequilibrium also important in circumstances in which there is a single equilibrium? Might disequilibrium be more likely the more nonstationary the world, and the faster the turnover in the population of agents playing the particular game? On the other extreme, is the limit of play relevant, and, hence, equilibrium coordination failure important in the macroeconomy?

Is the dynamic induced by repetition relevant, and, hence, such "learning" important in the macroeconomy? The "learning," or at least systematically changing behavior, itself presents interesting puzzles. Such "learning" is, of course, typical in experiments. Nevertheless, a standard view of game theory cannot address the "learning" observed in experiments. The game, as presented to the player, is a complete description of the relevant environment (supposedly), so there is nothing to be learned by the repetition. Of course, this standard view is a denial of the existence of strategic uncertainty. But why, then, do agents appear to "learn" in experiments? From a more practical perspective, experiments are, after all, hardly the subjects' first encounter with a social experience!

Perhaps experimental games are totally unlike the world. If this is so, it at least raises some questions about the immediate applicability of the results to the economy. Perhaps the lab setting in the experimental games is unfamiliar. What we are observing, then, is the formation of norms of behavior, as we might observe in any new situation (John Van Huyck has suggested this to me). This does, at least, however, suggest sensitivity to the setting. Perhaps behavior is very game specific. If so, the results may not be immediately applicable to the economy, if, perhaps, qualitatively interesting. Perhaps behavior is very group specific, although these particular experiments suggest otherwise. Perhaps some inate watchfulness suggests to people that they cannot be certain that lessons from one social situation will transfer whole, without amendment, to a somewhat different one. Perhaps "learning" is just the way people behave under strategic uncertainty. Certainly, as argued above, a standard view of game theory would give people no directions as to how to act when faced with other agents who are "learning!"

VHBB's experimental results raise the issue of the relative importance of extrapolation versus conceptualization in learning. This is most clearly seen in the aforementioned "overshooting." VHBB attribute the overshooting to conceptualizing, to subjects thinking about and learning the nature of the problem. Note that the "overshooting" cannot be explained by simple extrapolation. Crawford, on the other hand, treats what VHBB think of as conceptualization as a sort of inept or faulty extrapolation (this may not be Crawford's view of his approach). Crawford's approach does a remarkable job of tracking behavior in these, as well as other, experiments. Crawford's modeling of the observed dynamics is exciting, and may have many useful applications in economics, and elsewhere. This author does still wonder a little whether all aspects of the subject's learning by conceptualizing, as VHBB describe, are quite fully captured in Crawford's approach. The difference between conceptualization and extrapolation in learning may also be important in other Macroeconomic contexts. For example, in some overlapping generations models, extrapolative "learning" has agents converge to a "rational expectations" equilibrium which involves, in fact, the counter-factual belief that the present depends upon the past. In fact, the relevant equations are forward looking in these models. Truly conceptualizing agents might not converge to this counter-factual "rational expectation."

It may be appropriate at this time to congratulate, once again, VHBB on their work. The whole, totally new, dynamic that they brought to light, which we have been discussing, would not have been observed had VHBB used the two strategy format so often popular in experiments. We would have missed the striking sight of these model economies in a bottle, as it were, cascading downward. It may be, in the real economy, that at different levels of activity the character of the game changes locally, causing interesting alterations in the dynamic. VHBB, however, appropriately kept their game uniform so an unchanged dynamic could be observed.

Comparison of experimental results with the stag hunt and median rule pure coordination games reinforces some of the issues introduced previously. VHBB (1991) present experimental results on a pure coordination game which differs from the stag hunt in that subjects are "punished" for deviating from the group median action, not the group minimum action, as in the stag hunt. VHBB found, in the median rule experiments, that in the first play actions were widely dispersed. Moreover, the median action varied from experiment to experiment. Under repetition with the same group of subjects, the actions converged systematically to that particular group's original median action.

A standard view of game theory cannot address the differences between the experimental results in these two variants of pure coordination games. Both games involve strictly Pareto ranked Nash equilibria, and are, hence (supposedly), qualitatively the same. Perhaps it is strategic uncertainty, the existence of which is ignored by the standard view, which is the engine driving the differences in these experimental results. The minimum rule rewards pessimists, and it is the strategic uncertainty which generates a dispersion and, hence, optimists and pessimists. The rewarding of pessimists apparently then drives the downward dynamic of the stag hunt. (Parenthetically, Thomas Saving stressed the minimum rule's role in rewarding pessimists, if my memory serves me right.) In any case, these experimental results (as well as others) support the importance of off diagonal terms, and hence support the possible role of coordination theory in explaining why turning points are so hard to predict. The differences between the experimental results in these two variants of pure coordination games may also point to a theory of what structures are amenable to coordination, and by what institutions. It may also be that in a median type situation, one waits for a bad outcome before incurring the cost of setting up coordinating institutions, while with a minimum type (stag hunt) situation one gets right to it.

6. The Next Step?

This paper ends with a recommendation for a program of research. Experimental results to date suggest directions for further experimental research. Taken as whole, the experimental coordination game literature, to date, may appear to be a bit scattershot (this observation is not in reference to VHBB!) Possibly this is because reasonable economic models map into rather quirky strategic games, and experimentalists feel unfortunately constrained to experiments with the appearance of immediate economic application. Perhaps what is needed is a systematic exploration at the strategic game level. Then behavior can be modeled at the strategic game level. After this, economic models can be mapped into strategic games, where the derived theory of behavior is ready and waiting.

Finally, the author observes that Raymond Battalio (personal conversation) does have a counter argument to this proposed program of research. Not only does Battalio see current experiments as pin holes in a large fabric of strategic game structures, he sees them as pin holes inside of pin holes—so much depends upon the detail of the actual experimental setting. There might, then, be no hope of getting a good enough fix on the whole fabric to generate a useful behavioral theory. Battalio suggests, instead, running an experiment for each economic model, and using the experimental result as predictor, thereby dispensing with a behavioral theory. This raises to the author' mind a question. If results are really so sensitive to the experimental setting, will we ever get an experimental setting that is correct for the application of a model to aspects of the real economy, and, further, how situation specific must such results be? But, perhaps these issues cannot be addressed in the abstract, perhaps we must just go out and try.

7. Concluding Remark

As David Kreps (1990) says, this is high-risk stuff.

Acknowledgments

The author wishes to thank Marc Dudey, Scott Freeman, Peter Hartley, Richard Startz, Lawrence Summers, John Van Huyck, and especially James Friedman for very helpful comments. All errors and oversights are my responsibility alone.

Notes

1. Crawford (1991) reserves the term *stag hunt* for the minimum rule game. Others have used the term more generally, see Fudenberg and Tirole (1991). This paper follows Crawford's usage. If one traces the term back to Rousseau's stag hunt parable, it is not clear (at least to the author) exactly what Rousseau had in mind. Indeed, Rousseau may have had Prisoners' Dilemma or Moral Hazard considerations in mind, rather than coordination considerations. In any case, stag hunt is, at a minimum, a catchy phrase. Frequently in game theory a two player, two strategy version of the game is considered. While this simple version captures the essence of the logical dilemma of the coordination game, it may have less relevance for Macroeconomics, is overly restrictive, does not allow for the possibility (and experimental reality) that large groups behave differently from small groups, and eliminates the sort of dynamic analyzed by Van Huyck, Battalio and Beil (1990).

2. For example, the strategy spaces imposed in Jones and Manuelli (1992) are physically impossible in the stag hunt production game itself.

References

Bryant, John. 1980. "Competitive Equilibrium with Price Setting Firms and Stochastic Demand." *International Economic Review* 21 (October): 619–626.

Bryant, John. 1982. "Perfection, the Infinite Horizon and Dominance." *Economics Letters* 10: 223–229.

Bryant, John. 1983. "A Simple Rational Expectations Keynes-Type Model." *Quarterly Journal of Economics* 98 (August): 525–528.

Bryant, John. 1984. "An example of a Dominance Approach to Rational Expectations." *Economics Letters* 16: 249–255.

Bryant, John. 1987. "The Paradox of Thrift, Liquidity Preference and Animal Spirits." *Econometrica* 55 (September): 1231–1235.

Bryant, John. 1992. "Banking and Coordination." *Journal of Money, Credit and Banking* 24 (November): 563–569.

Colander, David. 1986. *Macroeconomic Theory and Policy*. Chicago: Scott Foresman.

Colander, David. 1992. "Is New Keynesian Economics New?" Middlebury College.

Cooper, Russell, Douglas V. DeJong, Robert Forsythe, and Thomas W. Ross. 1990. "Selection Criteria in Coordination Games: Some Experimental Results." *American Economic Review* 80 (March): 218–233.

Cooper, Russell, Douglas V. DeJong, Robert Forsythe, and Thomas W. Ross. 1992. "Communication in Coordination Games." *Quarterly Journal of Economics* 107 (May): 739–771.

Cooper, Russell, and Andrew John. 1988. "Coordinating Coordination Failures in Keynesian Models." *The Quarterly Journal of Economics* 103 (August): 441–463.

Crawford, Vincent. 1991. "An 'Evolutionary' Interpretation of Van Huyck, Battalio and Beil's Experimental Results on Coordination." *Games and Economic Behavior* 3 (February): 25–59.

Durlauf, Steven N. 1989. "Locally Interacting Systems, Coordination Failure, and the Behavior of Aggregate Activity." Stanford University (October).

Fudenberg, Drew, and Jean Tirole. 1991. *Game Theory*. Cambridge, Mass.: MIT Press.

Guesnerie, Roger. 1992. "An Exploration of the Eductive Justifications of the Rational-Expectations Hypothesis." *American Economic Review* 82 (December): 1254–1278.

Jones, Larry E., and Rodolfo E. Manuelli. 1992. "The Coordination Problem and Equilibrium Theories of Recessions." *American Economic Review* 82 (June): 451–471.

Keynes, John Maynard. 1936. *The General Theory of Employment, Interest and Money*. New York: Harcourt, Brace & World.

Keynes, Milo. *Essays on John Maynard Keynes*, Cambridge, Cambridge University Press, 1975.

Kreps, David. 1990. *Game Theory and Economic Modeling*. Oxford: Oxford University Press.

Mankiw, N. Gregory, and David Romer. 1991. *Coordination Failures and Real Rigidities, New Keynesian Economics* Vol. II, *MIT Press Readings in Economics*. B. Friedman and L. Summers (eds.) Cambridge, Mass: MIT Press.

Milgrom, Paul, and John Roberts. 1990. "Rationalizability, Learning, and Equilibrium in Games with Strategic Complementarities." *Econometrica* 58 (November): 1255–1277.

O'Driscoll, Gerald P. 1977. *Economics as a Coordination Problem: The Contributions of Friedrich A. Hayek*. Kansas City: Sheed, Andrews and McMeel.

Phelps, Edmund S. 1990. *Seven Schools of Macroeconomic Thought*. New York: Oxford University Press.

Phelps, Edmund S. 1991. *Recent Developments in Macroeconomics, The International Library of Critical Writings in Economics*. London: Edward Elgar.

Roberts, John. 1987. "An Equilibrium Model with Involuntary Unemployment at Flexible, Competitive Prices and Wages." *American Economic Review* 77 (December): 856–874.

Tobin, James. 1982. "Money and Finance in the Macro-Economic Process." *Les Prix Nobel En 1981* Stockholm: The Nobel Foundation.

Van Huyck, John B. Raymond C. Battalio, and Richard O. Beil. 1990. "Tacit Coordination Games, Strategic Uncertainty, and Coordination Failure." *The American Economic Review* 80 (March): 234–248.

Van Huyck, John B. Raymond C. Battalio, and Richard O. Beil. 1991. "Strategic Uncertainty, Equilibrium Selection, and Coordination Failure in Average Opinion Games." *The Quarterly Journal of Economics* 106 (August): 885–910.

Subject index

Backward induction, 33–34
Bandwagons, 151–154, 162, 169, 177
Battle of the sexes, 7, 9, 14, 34–35, 130
Behavior strategy, 24–26, 31–32
Behavioral equilibrium theory, 80–82

Cheap talk, 13, 18, 33, 36–37, 59, 70, 133–135, 141–143
Common knowledge, 20–22
Complete information, 20
Conjectural variation, 107
Consistent behavioral theory, 11, 73, 76–83
Coordination,
   Competing preferences, 9, 51, 54, 139–142
   Failure, 130
   Pure, 5, 50, 54, 130, 137–139

Dominant strategy, 137
Duopoly, 117–119

Equilibrium,
   Nash, 9
Evolutionary games, 61
Extensive form, 18–23

Focal equilibria, 51, 71, 82, 140
Folk theorem for repeated games, 13, 14, 18, 42–44, 52, 103, 108, 119–122, 186
   Imperfect monitoring and, 44–46, 187, 193–194, 197
   Tariff policy and, 187, 194–197
Forward induction, 13, 18, 34–35, 132, 140

Game tree, 19, 27
Games,
   Common interests, 4
   Imperfect monitoring, 44–46
   Mixed interests, 4
   Strictly competitive, 4

Individual rationality, 42, 194
Information set, 22, 27

Macroeconomics, 209–211
   Coordination and, 211–219
Markov perfect equilibrium, 156, 168
Minimax payoff, 42
Mixed strategy, 23, 79

Nature (as player), 24
Nash equilibrium, 26, 95
   And rationality, 32, 37, 174
Node,
   Decision, 19
   Terminal, 19

Paraperfect equilibrium, 91, 94–97
Pareto dominance, 36, 60

Payoff function, 23
Perfect equilibrium, 30–31
Perfect information, 20
Perfect monitoring, 40
Perfect recall, 24
Preplay communication, see cheap talk
Prisoner's dilemma, 8–12, 93–94, 98, 185
Pure strategy, 23, 158

Rationalizable strategies and equilibrium, 18, 37–39, 70, 72, 75
Reaction function, 103–104, 107–108, 115
Refinements of Nash equilibrium (see also specific refinements), 18, 27–39, 158
Repeated games, 39–40
  Paths in, 43, 109

Sequential equilibrium, 31–32
Stability, 104
  general equilibrium, 3
Strategic form, 18, 23
Strategy, definition of, 12, 90–91, 99
Subgame, 27–29, 41
Subgame perfect equilibrium, 27–30, 42–44, 93, 105, 108, 170

Tariff policy, 14, 185, 191
Technology
  Adoption, 150–152, 155, 162–164
  Quality of standard adopted, 164–166
  Standardization, 149, 161–163
  Switching cost, 152, 171–174
Trembles, 30
Trigger strategies, 40–42, 93, 108–109

Walrasian tâtonnement, 3

Author index

Abreu, Dilip, 18, 43, 44, 91, 126, 187, 188
Agliardi, Elettra
Arthur, Brian, 182
Arrow, Kenneth J., 3
Aumann, Robert J., 18, 53, 55–59, 91, 104, 108

Banks, Jeffrey S., 86
Basu, Kaushik, 85
Battalio, Raymond, 54, 59, 130, 211, 212, 216, 219–223
Beil, R., 54, 59, 130, 211, 212, 216, 219–223
Berg, J.E., 144
Bernheim, B. Douglas, 18, 70, 73, 75, 85
Bertrand, Joseph, 107
Biglaiser, Gary, 11
Binmore, Ken, 18, 100
Blume, Andreas, 14
Bowley, Arthur L., 104, 107
Bryant, John, 14, 129, 211, 217

Cabral, Luis, 183
Cho, In-Koo, 85, 86
Colander, David, 208, 211, 217
Cooper, Russell, 13, 129, 130, 143, 216
Cournot, Antoine Augustin, 3, 104, 107, 118
Crawford, Vincent P., 6, 51–54, 59–63, 85, 216, 219–222

Daley, L.A., 144
DeJong, Douglas, 13, 130, 143
Diamond, Peter, 129
Dickhaut, J.W., 144
Dixit, Avinash, 143, 194
Durlauf, Steven M., 217

Farrell, Joseph, 13, 18, 83, 85, 134, 141–143, 166, 182–184
Fellner, William J., 104, 108, 118
Forsythe, Robert, 13, 130, 143
Friedman, James W., 13, 18, 52, 90–92, 104, 105, 108, 110, 117–119
Fudenberg, Drew, 18, 43, 46, 52, 86, 104, 108, 110, 119, 120, 186, 188, 193, 194, 197, 222

Green, Edward, 44, 186
Guesnerie, Roger, 208
Gul, Faruk, 85

Hahn, Frank H., 3
Haller, Hans, 6, 52, 53, 54
Harsanyi, John, 36, 85, 143
Hayek, Friedrich, 208
Heller, Walter P., 129

Isoda, Kazuo, 26

John, A., 129, 216
Jones, Larry E., 215, 222

Kalai, Ehud, 96, 99
Katz, Michael, 129, 166, 182
Kennan, J., 190
Keynes, John Maynard, 208, 209, 215
Keynes, Milo, 215
Klemperer, Paul
Kohlberg, Elon, 18, 85, 86, 143
Kreps, David M., 18, 85, 86, 91
Kuhn, Harold, 26

Levine, David, 186, 188, 193, 194, 197
Luce, R. Duncan, 18, 104

Malouf, M.W.K., 135, 144
Mankiw, N. Gregory, 208
Manuelli, Rodolfo, E., 215, 222
Marschak, Thomas, 91, 126
Maskin, Eric, 18, 43, 46, 52, 104, 109, 110, 119, 120, 186–188, 193, 194, 197
Matsui, A., 54, 55, 57–59
Matsushima, H., 188
Maynard Smith, John, 61
Mertens, Jean-François, 18, 85, 86, 143
Milgrom, Paul, 91
Myerson, Roger, 18, 85, 187

Nash, John F., Jr., 26
Nikaido, Hukukane, 26

O'Brien, J.R., 144
O'Driscoll, Gerald P., 208

Pearce, David, 18, 44, 70, 73, 75, 85, 187
Phelps, Edmund S., 215
Porter, Robert, 44, 186

Rabin, Matthew, 11, 59, 83, 85, 86
Radner, Roy, 187
Raiffa, Howard, 18, 104
Riezman, Raymond, 14, 186, 190, 195
Roberts, John, 91, 208, 215

Robson, Arthur, 104, 108
Romer, David, 208
Ross, Thomas, 13, 129, 130, 143
Roth, Alvin E., 135, 144
Rousseau, Jean-Jacques, 14
Rubinstein, Ariel, 18, 90–92, 100, 104, 108, 119, 187, 188

Saloner, Garth, 143, 166, 182–184
Samet, Dov, 96
Samuelson, Larry, 12, 13, 105, 110, 117–119
Saving, Thomas, 222
Schelling, Thomas C., 143
Selten, Reinhard, 18, 36, 85, 86, 91, 126, 143
Shapiro, Carl, 13, 129, 143, 166, 182
Shapley, Lloyd, 18
Shubik, Martin, 90
Smith, Adam, 3
Sobel, Joel, 85, 86
Solow, Robert, 214
Sorin, Sylvain, 53, 55–58
Spence, A. Michael, 59
Stacchetti, Ennio, 44, 187
Stanford, William, 12, 96, 99, 100, 104, 108, 110, 117

Tirole, Jean, 18, 223
Tobin, James, 210

van Damme, Eric, 18, 143
Van Huyck, J., 54, 59, 130, 211, 212, 216, 219–223

Walras, Léon, 3
Watson, Joel, 85, 86
Weibull, Jorgen W., 85
Wilson, Robert, 18, 85, 91

Zapater, Inigo, 59